D0768848

San Diego Christian College
2100 Greenfield Drive
El Cajon, CA 92019

Undergraduate Texts in Mathematics

Editors

S. Axler
F. W. Gehring
K. A. Ribet

Springer
New York
Berlin
Heidelberg
Barcelona
Hong Kong
London
Milan
Paris
Singapore
Tokyo

Undergraduate Texts in Mathematics

Abbott: Understanding Analysis.

Anglin: Mathematics: A Concise History and Philosophy.
Readings in Mathematics.

Anglin/Lambek: The Heritage of Thales.
Readings in Mathematics.

Apostol: Introduction to Analytic Number Theory. Second edition.

Armstrong: Basic Topology.

Armstrong: Groups and Symmetry.

Axler: Linear Algebra Done Right. Second edition.

Beardon: Limits: A New Approach to Real Analysis.

Bak/Newman: Complex Analysis. Second edition.

Banchoff/Wermer: Linear Algebra Through Geometry. Second edition.

Berberian: A First Course in Real Analysis.

Bix: Conics and Cubics: A Concrete Introduction to Algebraic Curves.

Brémaud: An Introduction to Probabilistic Modeling.

Bressoud: Factorization and Primality Testing.

Bressoud: Second Year Calculus.
Readings in Mathematics.

Brickman: Mathematical Introduction to Linear Programming and Game Theory.

Browder: Mathematical Analysis: An Introduction.

Buchmann: Introduction to Cryptography.

Buskes/van Rooij: Topological Spaces: From Distance to Neighborhood.

Callahan: The Geometry of Spacetime: An Introduction to Special and General Relativity.

Carter/van Brunt: The Lebesgue–Stieltjes Integral: A Practical Introduction.

Cederberg: A Course in Modern Geometries. Second edition.

Childs: A Concrete Introduction to Higher Algebra. Second edition.

Chung: Elementary Probability Theory with Stochastic Processes. Third edition.

Cox/Little/O'Shea: Ideals, Varieties, and Algorithms. Second edition.

Croom: Basic Concepts of Algebraic Topology.

Curtis: Linear Algebra: An Introductory Approach. Fourth edition.

Devlin: The Joy of Sets: Fundamentals of Contemporary Set Theory. Second edition.

Dixmier: General Topology.

Driver: Why Math?

Ebbinghaus/Flum/Thomas: Mathematical Logic. Second edition.

Edgar: Measure, Topology, and Fractal Geometry.

Elaydi: An Introduction to Difference Equations. Second edition.

Exner: An Accompaniment to Higher Mathematics.

Exner: Inside Calculus.

Fine/Rosenberger: The Fundamental Theory of Algebra.

Fischer: Intermediate Real Analysis.

Flanigan/Kazdan: Calculus Two: Linear and Nonlinear Functions. Second edition.

Fleming: Functions of Several Variables. Second edition.

Foulds: Combinatorial Optimization for Undergraduates.

Foulds: Optimization Techniques: An Introduction.

Franklin: Methods of Mathematical Economics.

Frazier: An Introduction to Wavelets Through Linear Algebra.

Gamelin: Complex Analysis.

Gordon: Discrete Probability.

Hairer/Wanner: Analysis by Its History.
Readings in Mathematics.

(continued after index)

Peter Hilton
Derek Holton
Jean Pedersen

Mathematical Vistas

From a Room with Many Windows

With 162 illustrations

 Springer

Peter Hilton
Mathematical Sciences
  Department
SUNY at Binghamton
Binghamton, NY 13902
USA

Derek Holton
Department of Mathematics
  and Statistics
University of Otago
Dunedin
New Zealand

Jean Pedersen
Department of Mathematics
  and Computer Science
Santa Clara University
Santa Clara, CA 95053
USA

Mathematics Subject Classification (2000): 00A06, 00-01

Library of Congress Cataloging-in-Publication Data
Hilton, Peter John.
    Mathematical vistas : from a room with many windows / Peter
  Hilton, Derek Holton, Jean Pedersen.
        p.   cm. — (Undergraduate texts in mathematics)
    Includes bibliographical references and index.
    ISBN 0-387-95064-8
    1. Mathematics—Popular works.   I. Holton, Derek Allan, 1941–
  II. Pedersen, Jean.   III. Title.   IV. Series.
  QA93.H533   2002
  510 — dc21   00-056268

Printed on acid-free paper

Production managed by Steven Pisano; manufacturing supervised by Jerome Basma.
Composition and art by TEX Consultants, Three Rivers, CA.
Printed and bound by Hamilton Printing, Co., Rensselaer, NY.
Printed in the United States of America.

9 8 7 6 5 4 3 2 1

ISBN 0-387-95064-8   SPIN 10770592

Springer-Verlag   New York Berlin Heidelberg
A member of BertelsmannSpringer Science+Business Media GmbH

*A further tribute to those who showed us
how enjoyable mathematics can be,
especially as one penetrates deeper.*

Preface

Focusing Your Attention

We have called this book *Mathematical Vistas* because we have already published a companion book *Mathematical Reflections* in the same series;[1] indeed, the two books are dedicated to the same principal purpose — to stimulate the interest of bright people in mathematics. It is not our intention in writing this book to make the earlier book a prerequisite, but it is, of course, natural that this book should contain several references to its predecessor. This is especially — but not uniquely — true of Chapters 3, 4, and 6, which may be regarded as advanced versions of the corresponding chapters in *Mathematical Reflections*.

Like its predecessor, the present work consists of nine chapters, each devoted to a lively mathematical topic, and each capable, in principle, of being read independently of the other chapters.[2] Thus this is not a text which — as is the intention of most standard treatments of mathematical topics — builds systematically on certain common themes as one proceeds

---

[1]*Mathematical Reflections — In a Room with Many Mirrors*, Springer Undergraduate Texts in Mathematics, 1996; Second Printing 1998. We will refer to this simply as *MR*.

[2]There was an exception in *MR*; Chapter 9 was concerned with our thoughts on the doing and teaching of mathematics at the undergraduate level.

systematically through the book from Chapter 1 to Chapter 9. Any chapter that takes the reader's fancy may be studied at any time; and one is as likely to find in Chapter 4 a key reference to Chapter 8 as the other way round.[3]

Since the chapters are quite distinct from each other — though interrelated as the parts of mathematics inevitably are — each carries its own list of references.[4] We believe it will thus be clear to the reader why there is not simply a common list of references at the end of our text, but there is a common index.

We have chosen the topics treated with several considerations in mind. We have, as we have said, always hoped to intrigue and inform the bright and curious reader; but we do not believe that all topics should be treated at the same depth and with the same thoroughness. At one end of the spectrum we have the content of Chapters 3, 4, and 6, where students may be expected to have the necessary background for a complete understanding,[5] and where the proofs themselves constitute, in our judgment, an essential part of the appeal of the topics treated. At the other end of the spectrum we have Chapter 2, on Fermat's Last Theorem. The amazing triumph of Andrew Wiles in proving a famous result conjectured over 300 years ago is now — and forever will remain — a notable event in the history of the mathematical sciences. A book designed to highlight the attractions of mathematics should obviously tell the reader about this event, but it would be quite unrealistic to expect our readers to have the necessary technical background to be able to follow every step of the subtle argument used by Wiles and his colleagues. So we have discussed the history of FLT[6] in some detail, and sketched the arguments used in its proof.

Another consequence of the fact that the material of this book is not sequential, but rather consists of nine related mathematical essays, is that figures are sometimes repeated. We have preferred to repeat a figure rather than oblige our readers to study a figure that may be in a part of the book far from the pages currently being read.

We have thought of Chapters 1 and 2 as appetizers, though of a very different kind. Chapter 1 (Paradoxes) is quite self-contained and quite elementary; but it does serve to show that it is not sufficient to *learn* mathematics — one must also *think* about it. Unfortunately, many students of

---

[3] Actually — and exceptionally — we *do* recommend that Chapter 4 be read before Chapter 8, because Chapter 4 provides so much of the raw material for the examples used in Chapter 8.

[4] We include, in some cases, references to the web, which most of you will readily recognize from the format.

[5] Such a background is provided by a reading of Chapters 3, 4, 6, respectively, of *MR*.

[6] As Fermat's Last Theorem is abbreviated by all mathematicians.

mathematics today seem to believe it suffices just to commit mathematical techniques and arguments to memory, and regard it as almost a breach of contract if they are asked actually to reason mathematically. So Chapter 1 is included to help the reader to appreciate the proper role of thinking in doing mathematics, at any level.

Chapter 2 (FLT) is in our book to show how remarkable and rewarding mathematics is; but, of course, as we have said, it does not claim to be a complete treatment of its subject matter. The remaining chapters, however, are comprehensive in that they give clear[7] indications of all relevant arguments. Where we believe that the arguments or concepts are really difficult to grasp we have *starred* the material, that is, we have drawn a wavy line in the left-hand margin, with its initial and terminal points marked by a star. Sometimes an entire section has been starred. Starred material may, of course, be omitted on first reading if preferred.

Let us now describe some special features of our text. We have included a number of *BREAKS* in each chapter. These breaks consist of problems designed to enable our readers to test their understanding of the material thus far in the chapter. Answers to some of the problems appear at the end of the book. A complete set of answers to the problems in the breaks is available on request from Jean Pedersen, Department of Mathematics and Computer Science, Santa Clara University, Santa Clara, CA 95053.

Items in each chapter are numbered consecutively through the chapter. If an item in one chapter is referred to in another, then the chapter number is also given in the reference. We believe that the only conceivable confusion the reader may find in our numbering system is that n may refer either to formula (n) in the chapter or to the nth question of a given break. However, we have tried to ensure that the context always makes clear to which we are referring.

We adopt the standard abbreviations RHS and LHS for the *right-hand side* and *left-hand side*, respectively, of an equation, and the standard tombstone symbol □ to mark the end of a proof.

In connection with reading proofs, we advise the reader to follow, where appropriate, one of our guidelines (in Chapter 9 of *MR*), and to look at a *particular but not special* case; that is, to replace one of the parameters appearing in the argument by a particular, but not special, numerical value. By substituting a *particular* value the reader may well make the argument easier to understand; by avoiding a *special* value, the reader does not risk failing to recognize the true nature of the argument.

---

[7]We hope!

The authors wish to thank Dr. Ina Lindemann very warmly for her encouragement and very valuable cooperation during the production of our manuscript, and to express to the referee their gratitude for his very helpful and expert comments on an earlier version of Chapter 2, which enabled us to improve our presentation substantially. The authors also wish to extend their grateful thanks to Kent Pedersen and Roberta Telles for the essential help they provided in assembling the Index. Further, we have benefited greatly from the collaboration of Hans Walser in the preparation of Chapter 6, and from consultation with Robert Bekes, John Holdsworth, Peter Ross, and Richard Scott.

Contents

Preface: Focusing Your Attention **vii**

1 **Paradoxes in Mathematics** **1**

 1.1 Introduction: Don't Believe Everything You See and Hear . 1

 1.2 Are Things Equal to the Same Thing Equal to One
Another? (Paradox 1) 4

 1.3 Is One Student Better Than Another? (Paradox 2) 6

 1.4 Do Averages Measure Prowess? (Paradox 3) 8

 1.5 May Procedures Be Justified Exclusively by Statistical
Tests? (Paradox 4) 11

 1.6 A Basic Misunderstanding — and a Salutary Paradox
About Sailors and Monkeys (Paradox 5) 14

   References 20

2 **Not the Last of Fermat** **23**

 2.1 Introduction: Fermat's Last Theorem (FLT) 23

 2.2 Something Completely Different 24

 2.3 Diophantus 26

 2.4 Enter Pierre de Fermat 27

 2.5 Flashback to Pythagoras 28

2.6   Scribbles in Margins   32

2.7 $n = 4$ 33

2.8   Euler Enters the Fray   36

2.9   I Had to Solve It   40

     References   46

3   Fibonacci and Lucas Numbers: Their Connections and Divisibility Properties   49

3.1   Introduction: A Number Trick and Its Explanation   49

3.2   A First Set of Results on the Fibonacci and Lucas Indices .   54

3.3   On Odd Lucasian Numbers   56

3.4   A Theorem on Least Common Multiples   62

3.5   The Relation Between the Fibonacci and Lucas Indices . .   63

3.6   On Polynomial Identities Relating Fibonacci and Lucas Numbers   64

     References   69

4   Paper-Folding, Polyhedra-Building, and Number Theory   71

4.1   Introduction: Forging the Link Between Geometric Practice and Mathematical Theory   71

4.2   What Can Be Done Without Euclidean Tools   73

4.3   Constructing All Quasi-Regular Polygons   93

4.4   How to Build Some Polyhedra (Hands-On Activities) . . .   95

4.5   The General Quasi-Order Theorem 114

     References 124

5   Are Four Colors Really Enough?   127

5.1   Introduction: A Schoolboy Invention 127

5.2   The Four-Color Problem 127

5.3   Graphs 130

5.4   Touring with Euler 136

5.5   Why Graphs? 138

5.6   Another Concept 142

5.7   Planarity 144

5.8   The End 148

5.9   Coloring Edges 149

5.10   A Beginning? 153

       References 157

6   From Binomial to Trinomial Coefficients and Beyond   159

6.1   Introduction and Warm-Up 159

6.2   Analogues of the Generalized Star of David Theorems . . 177

6.3   Extending the Pascal Tetrahedron and the
       Pascal *m*-simplex 188

6.4   Some Variants and Generalizations 190

6.5   The Geometry of the 3-Dimensional Analogue of the
       Pascal Hexagon 193

       References 198

7   Catalan Numbers   199

7.1   Introduction: Three Ideas About the Same Mathematics . . 199

7.2   A Fourth Interpretation 208

7.3   Catalan Numbers 215

7.4   Extending the Binomial Coefficients 218

7.5   Calculating Generalized Catalan Numbers 220

7.6   Counting *p*-Good Paths 223

7.7   A Fantasy — and the Awakening 227

       References 233

8   Symmetry   235

8.1   Introduction: A Really Big Idea 235

8.2   Symmetry in Geometry 239

8.3   Homologues 254

8.4   The Pólya Enumeration Theorem 257

8.5   Even and Odd Permutations 263

       References 269

9   Parties   271

9.1   Introduction: Cliques and Anticliques 271

9.2   Ramsey and Erdős 275

9.3   Further Progress 277

9.4 $N(r, r)$ 281

9.5   Even More Ramsey 283

9.6   Birthdays and Coincidences 285

9.7   Come to the Dance 287

9.8   Philip Hall 290

9.9   Back to Graphs 292

9.10 Epilogue 295

      References 297

Selected Answers to Breaks   **299**

Index   **325**

1

CHAPTER

Paradoxes in Mathematics

1.1 INTRODUCTION: DON'T BELIEVE EVERYTHING YOU SEE AND HEAR

A good dictionary will give two (or perhaps more) distinct meanings of the word "paradox." The first meaning is "a self-contradictory statement"; thus the celebrated classical paradox due to Bertrand Russell (1872–1970), based on a naive set theory which permits the proposition $x \in x$. A popular, but not very accurate, form of this paradox is contained in the following little story:

> In the little village of Humblemeir there is a male barber. Now this barber shaves only those men who don't shave themselves. When you think about it, this leads to a paradox. Who shaves the barber? If he shaves himself, he doesn't, because he only shaves those who don't shave themselves. On the other hand, if he doesn't shave himself, he does, because he only shaves those who don't shave themselves.

So we apparently have a paradox. But, if you think deeper about it, you see that the statement "...the barber shaves *only* those..." allows the possibility that the barber never has a shave, since it does not claim that the barber shaves *everybody* who doesn't shave himself.[1]

---

[1]For those who are comfortable with set theory, here is the formally accurate form of the paradox. The paradox goes as follows. Let s be the set of all x such that $x \notin x$. We conclude that

We cannot tolerate such paradoxes as the Russell paradox in mathematics (though a celebrated theorem of Gödel tells us we cannot guarantee that our mathematical system is free of them); if we do meet such a paradox we must modify our logical axioms to eliminate it.

However, we are concerned in this chapter with a different use of the term "paradox"; this refers to an assertion that is an affront to conventional or accepted thinking. Thus the famous writer Bernard Shaw, defending the introduction of a National Health Service in Great Britain after the Second World War, wrote, "My main reason for supporting its introduction is that it interferes with the freedom of the doctor." While most apologists for the NHS had been at great pains to argue that *no* curtailment of the freedom of the doctor would result, Shaw, a life-long critic of the medical profession, outraged conventional opinion by implying that such a curtailment would be a thoroughly good thing. Again, the celebrated wit Oscar Wilde was responsible for the aphorism "Work is the curse of the drinking classes." By transposing the roles of work and drink, Wilde stood conventional morality on its head.

Mathematical reasoning may well lead to such paradoxes.[2] If it does so, there is no implied criticism of the mathematics — it is the conventional reasoning that is called into question by the paradox. We do not need to repair our mathematical system, since no contradiction has been revealed within the system. We do need to revise our conceptual thinking in order to avoid a popular misconception which our mathematics has uncovered.

We propose in this chapter to discuss five examples of such paradoxes. We do not claim originality for any of them, although, as a matter of fact, we have not seen a paradox of the fourth kind in the literature. Each paradox is introduced by a description of the popular view that is affronted by the paradox and accompanied by a suggestion as to how the paradox may be avoided.

We argue that the study of paradoxes of this kind has an important role (or, perhaps, several important roles) to play in education. In these days when we are constantly being told by the media, the politician, and the commercial advertiser what we should think and believe, it is more than ever necessary to adopt a position of healthy skepticism with regard to claims and assertions that are presented to us as self-evident truths. It is very enlightening to see that mathematical reasoning can enable us to

$$\text{if } s \in s, \text{ then } s \notin s,$$
$$\text{and if } s \notin s, \text{ then } s \in s.$$

Here, of course, we have been led to an actual contradiction.

[2]For further examples of interesting paradoxes see [5].

discover weaknesses and inconsistencies in conventional viewpoints. Let us give just one illustration of this point here in the introduction.

In many of our activities we are faced with tables in which members of a set are ranked according to some attribute or property. We place baseball players in order according to their batting averages, we place students in order according to some measure of their performance in an examination. Now such ranking presupposes an ordering that is *transitive* and *linear*. Here, an ordering of a set S is said to be **transitive** if, whenever $a < b$ and $b < c$, we may infer that $a < c$; and it is said to be **linear** or **total** if, for any elements a, b in S, we have $a < b$, $a = b$, or $a > b$. Thus the set of positive integers is ordered by **size**, when $a < b$ means that a is a smaller integer than b; but it may also be ordered by **divisibility**, where $a < b$ means that $a|b$. The first ordering is total, but the second is not. (Why not?) An ordering that is not linear (i.e., not total) is usually called **partial**.

Now, if an attribute is multidimensional, the ordering fails to be linear — there is no natural linear ordering of the points in the plane, for example, even if we confine ourselves to points with integer coordinates.[3] Moreover, for more complicated situations, the relation in question may even fail to be transitive. Where a sophisticated skill is involved, of the kind we test in a math exam, it may well happen that A usually beats B, and B usually beats C, without it being true that A usually beats C. Thus such ranking tables implicitly contain certain assumptions that our mathematical training should cause us to question.[4] Without this skepticism we are led to assume unquestioningly that A is better than B if A appears higher than B on our list. The very use of a comparative adjective "better" presupposes the presence of a *transitive* relation and thus biases our thinking. We view this age-old error with regard to the ranking of students as a pernicious aspect of traditional education. There are many qualities involved in doing mathematics successfully (skill, imagination, accuracy, logical thinking, speed of execution, to name some), so that a measure of mathematical success should be a multidimensional vector; and we do much harm by crudely and arbitrarily converting that vector, for administrative convenience, into a scalar. We may describe this as superimposing an avoidable error — linear ordering — on an unavoidable error — the inherent unreliability of tests.

However, paradoxes of this kind, conflicting with conventional assumptions and ways of thinking, need not arise from a mathematical argument

---

[3]Of course, you could define a linear ordering, since there is a one-to-one correspondence between points in the plane and the set of real numbers, but the resulting ordering would not be natural, in the sense of fitting well with the algebra of the real numbers.

[4]Many of the paradoxes in this chapter illustrate failure of transitivity in what appears to be a ranking situation.

directly. There are paradoxes arising from the nature of mathematical reasoning itself, and we would like to deal with one of the most pervasive of these as our final example in this chapter.

Many people argue that since, in teaching mathematics, we want to enable our students to use mathematics in the real world, we must make the mathematics more *concrete* and less *abstract*. We have no quarrel with the viewpoint that, for most students, the objective of being able to apply mathematics to a real-world problem is paramount,[5] but that is no reason to try to make the mathematics concrete in order to match that world. For mathematics is, by its very nature, abstract — numbers do not exist in the real world but are *abstracted* from our real-world experience of numbers of apples, numbers of chairs, numbers of students, ... Thus the paradox is that, to draw conclusions about the real world, we construct an abstract, mathematical model that is *not* a part of the real world. What's more, we reason within that model, using concepts and constructions that may have no meaning in the real-world context. This remarkable feature of mathematical reasoning, then, is the topic of our fifth example. Each of the following five sections of this chapter will be devoted to a discussion of a particular paradox.

● ● ●   BREAK 1

Here is a paradox for you to think about. We ask four questions.

— What is the probability that an integer, chosen at random, is divisible by 2?
— What is the probability that an integer, chosen at random, is divisible by 3?
— Does the function $2n \mapsto 3n$ set up a one-to-one correspondence between integers divisible by 2 and integers divisible by 3?
— How, then, can the probabilities be different?

1.2   ARE THINGS EQUAL TO THE SAME THING EQUAL TO ONE ANOTHER? (PARADOX 1)

The answer is yes, if we speak of equality as identity. Thus, in set theory, we write $A = B$ only if A and B are the same (numbers, elements, or sets), and then we may well conclude that if $A = B$ and $B = C$, then $A = C$, since the notion of sameness is surely transitive. But in our daily lives we may well describe A and B as equal if each is equally likely

---

[5]Though we would caution that it must be the world of the student, not the world of the (adult) teacher.

to win in a contest between the two. That this is a dangerous use of the word "equal" we demonstrate by the paradox in this section, where it is shown that this sort of equality is not transitive. The danger is removed by recognizing that conventional usages do not always correspond to precise mathematical usages, and that we should therefore beware of believing ourselves to be in the presence of an equivalence relation (in particular, a *transitive* relation) whenever equality is mentioned in ordinary day-to-day conversation.

Consider this example: A machine is producing a sequence of 0's and 1's at random. Let this sequence for a given "game" be S. Suppose that

> player A is assigned the ordered pair $(0, 1)$;
> player B is assigned the ordered pair $(1, 0)$;
and player C is assigned the ordered pair $(1, 1)$.

Then, if A is playing against B, A wins if, in the sequence S, the pair $(0, 1)$ occurs before the pair $(1, 0)$. Similarly, if A plays against C, A wins if, in the sequence S, the pair $(0, 1)$ occurs before the pair $(1, 1)$. A similar rule applies when B plays against C. For example, if the sequence S were $0010110\cdots$, then A would beat B (since $(0, 1)$ starts in place 2, while $(1, 0)$ starts in place 3), B would beat C, and A would beat C. Of course, we expect transitivity in a *single* game, since the notion of "before" is certainly transitive. However, let us consider which of A and B is *more likely* to win the game. We claim that, by reasons of symmetry, they must be equally likely to win, since the machine is producing the sequence at random. Thus we might write $A = B$ (but remember that this is a dangerous notation).

We also claim that $B = C$. In this case we argue as follows. Neither B nor C is interested until the machine produces a 1. Then both become interested, and the issue is decided by the next digit produced by the machine; if it is 0, B wins, and, if it is 1, C wins.

However, we find a very different situation if A plays against C. Suppose that the machine initially produces 0. We claim that then A *cannot lose*. For we simply have to wait till the machine produces a 1 for A to achieve the sequence $(0, 1)$; C never has a chance. Now suppose that the machine first produces 1; in that case we consider the next number it produces. If the next number is 1, then, of course, C wins. But if the next number is 0, the situation once again is that A cannot lose. Thus the probabilities may easily be computed from a tree diagram. (See Figure 1 below.)

We find that the probability that A beats C is $\frac{3}{4}$. Thus, $A = B$, $B = C$, but $A > C$, meaning that in a contest between A and C, A is likely to win; indeed, the odds on A are 3 to 1.

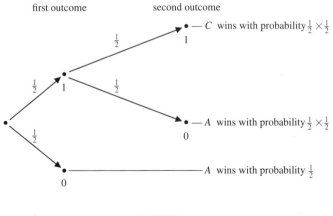

FIGURE 1

● ● ● BREAK 2

Now suppose that player A is assigned the ordered triple $(0, 0, 1)$, player B is assigned the ordered triple $(1, 1, 0)$, and player C is assigned the ordered triple $(1, 1, 1)$. Show that $A = B$, $B = C$, but $A > C$. (Hint: The subtle part is working out the probabilities when A plays C, as you might expect.)

1.3   IS ONE STUDENT BETTER THAN ANOTHER? (PARADOX 2)[6]

You may think that this is true of students, but the simple example below shows that it is not true of certain crazy dice! Here we understand a crazy die[7] to be a cube, each of whose faces is labeled by an integer, but not in the usual way. If dice A and B are thrown, we say that A wins if the top face of A is labeled with a number that exceeds that on the top face of B. It is thus reasonable to ask whether A is likely to beat B, and, if so, what is the probability that A beats B. It is common parlance to describe A as better than B if this probability exceeds $\frac{1}{2}$; and it is in the unthinking adoption of this terminology, involving a comparative adjective, that the error lies. Tennis player A may indeed be better than player B; but, if our

---

[6]A more sophisticated version of this type of paradox is found in [6]. Somewhat different versions of the paradox are to be found in [8, 9]. An algorithm for labeling the faces of other Platonic solids (except the tetrahedron, for obvious reasons) to produce similar paradoxes can be found in [10].

[7]These days, many people would say "a crazy dice," but this is not really correct, since "dice" is the plural form.

evidence is simply A's tendency to beat B, then we are — as our example will show — in danger of having to say that A is better than B, B is better than C, and C is better than A. And if this is true of tennis players, is it not also true of students?

Let the three crazy dice A, B, C have their six faces labeled as shown below:

A	5	5	5	5	1	1
B	4	4	4	4	4	4
C	6	6	2	2	2	2

A pair of dice is rolled, and the one showing the highest number on the top is said to be the winner.

It should be clear that A beats B with a probability of $\frac{2}{3}$, so that we may write $A > B$, "A is better than B." Likewise, B beats C with a probability of $\frac{2}{3}$, so that we may write $B > C$, "B is better than C." Now consider a match between A and C; here the situation is not quite so obvious, but a simple tree diagram allows us to compute the probability. (See Figure 2.) Thus the probability that C wins is $\frac{2}{9} + \frac{1}{3} = \frac{5}{9}$, so C beats A with a probability of $\frac{5}{9}$, so that we may write $C > A$, usually described by the phrase "C is better than A." We conclude that $A > B > C > A$; it is then pertinent to ask "Which die would you rather have?" We leave the answer to the reader, but offer the remark that this may not be the same question as "Which die is best?" At the very least, we should stop saying "A is better than B," replacing it by "A usually beats B."

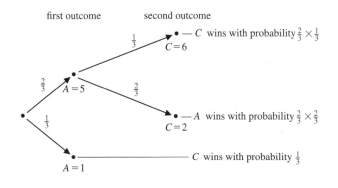

(We suppose A is rolled first)

FIGURE 2

● ● ●   **BREAK 3**

(1) Can you label the six faces of four dice A, B, C, D in such a way that, in the sense used above, $A > B > C > D > A$?

(2) Which of the dice in part (1) would you rather have? Or doesn't it matter?

(3) Can a similar situation be created for 5 dice?

1.4   DO AVERAGES MEASURE PROWESS? (PARADOX 3)

In many sports the standing of an individual player, or of a team, is measured by the ratio of the number of successful attempts to the total number of attempts. For example, simplifying slightly, a baseball player's average is computed by dividing the number of hits by the total number of occasions at bat during the specified time interval. This is called the player's batting average over that time interval.[8] Tables are provided at regular intervals listing the batting averages of the leading players (in decimal form) up to the day in question. Moreover, the players are listed in order of decreasing batting average, thus encouraging the popular view that player A is better than player B if player A has a higher batting average. This view is logically untenable, as the following example shows.

Suppose that in period I player A has 3 successes in 4 attempts and player B has 2 successes in 3 attempts; and that in period II player A has 4 successes in 11 attempts while player B has 1 success in 3 attempts. If the entire season consists of periods I and II, then Figure 3 displays the players' averages during each period separately and over the entire season.

We now notice that since $\frac{3}{4}$ ($= 0.750$) is greater than $\frac{2}{3}$ (≈ 0.667), and $\frac{4}{11}$ (≈ 0.364) is greater than $\frac{1}{3}$ (≈ 0.333), player A is judged to have played better than player B during both periods I and II; but since $\frac{3}{6}$ ($= 0.500$)

Player	Average for		
	Period I	Period II	Entire season
A	$\frac{3}{4} = 0.750$	$\frac{4}{11} \approx 0.364$	$\frac{7}{15} \approx 0.467$
B	$\frac{2}{3} \approx 0.667$	$\frac{1}{3} \approx 0.333$	$\frac{3}{6} = 0.500$

FIGURE 3

[8]Of course, a similar average is calculated for a batsman in cricket.

is greater than $\frac{7}{15}$ ($\approx$ 0.467), we are forced to conclude, by the same criterion, that player B played better than A over the entire season! Such a set of conclusions, taken together, makes obvious rubbish. We can avoid the paradox by abandoning the view that the average is a sensible measure of a player's prowess. We remark that if instead we measured performance by the *difference* between the number of successes and the number of failures, then the anomaly revealed in our example could not occur. There may, of course, be other objections to this proposed procedure — we encourage the reader to consider what they might be.

A different analysis of the potential defects of measuring the prowess by means of averages may be found in [1].

The reader may believe that the paradox in our example arose because player A made many more attempts than player B. This is not so; we may consider two students who must take 10 pass-or-fail tests in the course of the semester. In the first half of the semester, student A took 8 tests and passed 2, while student B took 5 tests and passed 1; in the second half , student A took the 2 remaining tests and passed both, while student B took the 5 remaining tests and passed 4. See Figure 4.

We "conclude" that student A did better over each half-semester, but student B did better over the whole semester!

The perceptive reader will notice that, in this section, we have found ourselves adding fractions by an illegitimate method! The fact is that, *strictly speaking*, this is a perfectly valid method of combining fractions, but not of adding the rational numbers they represent. Since we use the fractions to represent numbers (in order to say who performed "better"), we shouldn't wonder that paradoxes arise.

Historical Note   Readers may be interested to know that instances of this paradox — known to many as ***Simpson's Paradox*** (see [7]) — have already been noted in the annals of United States major league

Student	Average for		
	First half	Second half	Semester
A	$\frac{2}{8}$	$\frac{2}{2}$	$\frac{4}{10}$
B	$\frac{1}{5}$	$\frac{4}{5}$	$\frac{5}{10}$

FIGURE 4

	Ken Oberkfell			Mike Scioscia		
	Hits	At-bats	Batting average	Hits	At-bats	Batting average
1983	143	488	.293	11	35	.314
1984	87	324	.269	93	341	.273
Combined (1983–84)	230	812	.283	104	376	.277

	Dave Justice			Andy Van Slyke		
	Hits	At-bats	Batting average	Hits	At-bats	Batting average
1989	12	51	.235	113	476	.237
1990	124	439	.282	140	493	.284
Combined (1989–90)	136	490	.278	253	969	.261

FIGURE 5

baseball. The table shown in Figure 5, published in the *American Mathematical Monthly* in 1992 (see [2]), compares, first, the averages of Ken Oberkfell and Mike Scioscia in 1983 and 1984, and, second, the averages of Dave Justice and Andy Van Slyke in 1989 and 1990. You will see that Oberkfell had an inferior average to Scioscia in each of the seasons 1983, 1984, but a superior average over the two seasons combined; likewise, Justice had an inferior average to Van Slyke in each of the seasons 1989, 1990, but a superior average over the two seasons combined.[9]

• • • BREAK 4

(1) Make up an example, similar to the one shown in Figure 4, but with more tests given (and not just multiples of the numbers in Figure 4!).

(2) Make up a baseball scenario where there are three parts to the season and such that the batting average of player *A* is higher than the batting average of player *B* in all three parts of the season, but player *B* has an overall batting average that is better than the overall batting average of player *A*.

---

[9]We are indebted to Dr. Peter Ross for drawing our attention to Friedlander's note [2] — and for many other helpful suggestions.

(3) Make up an example like that shown in Figure 4 where students *A* and *B* take the same number of tests over a semester, and the semester is divided into three parts.

1.5 MAY PROCEDURES BE JUSTIFIED EXCLUSIVELY BY STATISTICAL TESTS? (PARADOX 4)

Today data are collected with great speed and efficiency. We are bombarded with statistics; and we are encouraged to believe that firm conclusions may be drawn from those statistics ("eating butter predisposes you to cancer," "women are more prone than men to jet-lag").

Often the data take the form of a 2×2 table.[10] Thus a given treatment *T* is under consideration. A group of participants in the experiment is chosen; to some members of the group the treatment is administered, but not to the rest. A certain outcome is considered favorable, and we count the number of *favorable* outcomes (*F*) and *unfavorable* outcomes (*U*), both among those *treated* (*T*) and those *not treated* (*N*). The results are presented as a 2×2 table. (See Figure 6.)

It is plain that we would want to maximize the number of favorable outcomes. Hence if $\frac{a}{c}$ is significantly greater than $\frac{b}{d}$, that is, if *ad* is significantly greater than *bc*, then we conclude that the treatment is effective; in the contrary case we conclude that it is not.

The question of statistical significance is important here, but we need not concern ourselves with it in our discussion. For given entries *a*, *b*, *c*, *d* in our example, we may change the example to *ka*, *kb*, *kc*, *kd*, and by choosing *k* large enough, we will achieve statistical significance at any desired level. Thus we may always suppose our 2×2 tables statistically significant.

Now, we do not criticize the procedure of reasoning from 2×2 tables if a well-founded scientific hypothesis is being tested by experiment; such an experiment certainly serves to strengthen (or weaken, perhaps fatally)

	T	N
F	a	c
U	b	d

FIGURE 6

[10]This is usually read "two-by-two table."

our confidence in the hypothesis. But we do criticize a blind reliance on the statistics in the absence of a solid scientific basis of belief.

The following example shows why; you will see that, in presenting this example, we indulge in a little fantasy. Mathematics, after all, should appeal to and develop the imagination. We tell our fairy tale as follows.

Once upon a time in the peaceful country of *Borgia* the successful playing of Ping-Pong was considered to be a very prestigious activity. When a dispute arose with the neighboring warlike country of *Mackerelroe*, the Prime Minister of Borgia, on behalf of his loyal countrymen, challenged the Prime Minister of Mackerelroe to an enormous Ping-Pong tournament between the two countries.[11] Since Ping-Pong had been a revered, and hence skillful, pursuit of the citizens of both countries for thousands of years, the challenge was readily accepted and, as fate would have it, it was quickly discovered that each country had exactly 2,090,000 Ping-Pong players who had been granted "National Ping-Pong Player" status. The matches between players were arranged by a lottery and the date was set for the tournament.

At this stage, the Mackerelroe Minister of Health, Dr. Ignoramus Fuddle-Thought, popularly known as "Drift," intervened to propose that the tournament provided an excellent opportunity to test a potentially wonderful potion that he had himself invented and called ***Everswear***. He claimed that his potion would greatly increase a player's chances of success, and suggested that all the Mackerelroe players should drink his potion before their tournament match. Not all agreed — to his intense annoyance — but he determined to show the players just how beneficial his potion was.

In fact, Mackerelroe won the tournament 1,100,000 to 990,000; but Drift was more interested in demonstrating the effectiveness of the potion than in celebrating the victory. Accordingly, he recorded the results (in 10,000's) as follows, separating his players according to age and sex (see Figure 7).

Since, in every case, "$ad - bc > 0$," Drift was jubilant, announcing confidently that ***Everswear*** had contributed significantly to Mackerelroe's success.

However, these were days of great social upheaval in Mackerelroe. The Committee against Age Discrimination (CAD) objected to the filing of separate figures for players over and under 30. They therefore insisted on publishing, both for men and for women, the combined figures, irrespective

---

[11]Readers who are at least 30 years old will vividly recall the great feats of two outstanding tennis players, Bjorn Borg and John McEnroe, and their very different temperaments.

	Potion	No potion
M wins	6	8
B wins	6	9

Men over 30
$(6 \times 9 > 6 \times 8)$

	Potion	No potion
M wins	6	22
B wins	3	12

Men 30 or under
$(6 \times 12 > 3 \times 22)$

	Potion	No potion
M wins	4	17
B wins	6	27

Women over 30
$(4 \times 27 > 6 \times 17)$

	Potion	No potion
M wins	4	43
B wins	3	33

Women 30 or under
$(4 \times 33 > 3 \times 43)$

FIGURE 7

of age. These were (as you may verify from Figure 7) those given in Figure 8.

CAD duly reported, to Dr. Fuddle-Thought's horror, that, in fact, the potion disadvantaged both men and women players, since "$ad - bc < 0$" held for both men and women.

Reeling under this blow, Drift was next confronted with a delegation from the Association against Sexual Supremacy (ASS), who objected vehemently to the publication of separate figures for men and women. They insisted on amalgamating the two tables — a procedure to which Drift made no objection, expecting that his faith in **Everswear** would thereby be vindicated. The final table, then, was as shown in Figure 9.

Thus, since "$ad - bc = 0$," the entire experiment, involving 2,090,000 of Mackerelroe's most outstanding citizens, was totally inconclusive. The unfortunate Dr. Fuddle-Thought, unable to reconcile himself to his confu-

	Potion	No potion
M wins	12	30
B wins	9	21

Men
$(12 \times 21 < 9 \times 30)$

	Potion	No potion
M wins	8	60
B wins	9	60

Women
$(8 \times 60 < 9 \times 60)$

FIGURE 8

	Potion	No potion
M wins	20	90
B wins	18	81

All players
$$(20 \times 81 = 18 \times 90)$$

FIGURE 9

sion and disappointment, took a massive overdose of ***Everswear***, trampled underfoot his favorite Ping-Pong paddle,[12] let out a shrill and prolonged expletive, and expired.

The moral is clear. Statistics can, and should, be used to test a scientific hypothesis for which there is already some evidence and strong theoretical underpinning. In the absence of such a hypothesis they can be highly unreliable!

1.6   A BASIC MISUNDERSTANDING — AND A SALUTARY PARADOX ABOUT SAILORS AND MONKEYS (PARADOX 5)

We believe that there may well be prevalent even among many teachers of mathematics a serious misunderstanding about the nature of mathematics and about the implications, for the teaching of mathematics, of the dictum that we must make the teaching of mathematics relevant to the world of the student.

The misunderstanding, in its starkest form, consists in believing that, if the mathematics is to be relevant to real-world problems, then it must somehow take place itself in the real world; that is, it must be concrete, and not abstract. We must always deal directly with real-world objects; thus, mathematical "theory" is to be replaced by rules of calculation using hand-calculators, microcomputers, and computers, and we should be concerned to get numerical answers to specific problems rather than formulae of a general nature.

This viewpoint is grossly mistaken, so much so that we have even found it difficult to describe it in such a way that it could be held by any intelligent person. It is mistaken because it fails to recognize that *mathematics is, by its very nature, abstract*, whether or not the problem to be solved is a real-world problem or a problem within mathematics itself. When

---

[12] Some of our readers may call this a bat.

we reason mathematically, we construct, explicitly or implicitly, a mathematical model of the real-world situation,[13] and we *reason within the model*. This process of passing to the mathematical model is *mathematical abstraction*. It is a subtle process; it is an art and it cannot be made algorithmic. To be able to construct a mathematical model offering a reasonable expectation for success in solving a problem, one must understand the problem well and one must be able to reason effectively within the model. Thus non-mathematical intelligence, together with mathematical experience, understanding, and skill, are all vital ingredients of successful problem-solving.[14] If, as so often happens in traditional mathematics courses, the student is given the mathematical model, with no explanation of why it is appropriate to a certain class of problem and no discussion of its theoretical features, and is then asked to carry out some mechanical steps on the model with no real understanding of their validity and significance, then the student is indeed simply "manipulating symbols." But constructing a model, reasoning within the model, and then executing well-motivated calculations — these are the very stuff of mathematics itself.

A companion misunderstanding of the nature of mathematics is revealed by the bad advice, which one hears so often even from reputable sources, always to keep the real-world problem in mind when doing the mathematics. The philosopher Bertrand Russell once described mathematics as the subject "in which you don't know what you're talking about and don't care whether what you're saying is true." The intended meaning behind this deliberately provocative aphorism is that criteria of truth, in the sense of correspondence to reality, are inapplicable to an axiomatic system; and that the concepts of such a system are not themselves "real," however useful they may be to solving a real problem, and thus not "knowable" in the sense that an apple or a ladder is knowable.

No, the better advice is to trust the mathematical model, to stay within the model until the time is appropriate for checking the theoretical deductions and the calculations against the original real-world situation. There is, of course, some risk in this — one may not have made a clear statement of what information one is hoping to derive from the model, and thus one may go off on a wild goose chase of irrelevant reasoning and pointless calculation. But the remedy is to use mathematical reasoning more efficiently, not to clutter the model with incongruous concrete props.

---

[13]Of course, there may well be a *scientific model*, or *conceptual model*, intervening between the real-world problem and the mathematical model. (See [3].)

[14]We oppose the teaching of problem-solving as a separate skill — of course!

Let us give two very elementary examples here of this point we are making that the model is abstract and that we do mathematics with abstract concepts and entities — of which "number" is the most fundamental. Then we will be ready to give a more sustained and sophisticated example of the value and validity of abstract reasoning for solving concrete problems — this will be our "salutary paradox."

There is a popular phrase, "That's like adding apples and oranges." The thought behind this is that you can only add like objects. The truth of the matter is that you can only add *numbers*, if addition is the arithmetical operation familiar to all children. It is true that you can't add apples and oranges; but, strictly speaking, it is also true that you can't add apples and apples. To find out how many apples you have if, already possessing 6 apples, you are given 4 more, you add the *numbers* 6 and 4. This mathematical model is also appropriate for very different real-world situations — if you are 6 miles east of Chicago and travel a further 4 miles east, you add $6 + 4$ to find out how many miles you are now east of Chicago. Of course, once it is *understood* that only numbers can be added, we should allow ourselves to talk of adding 6 apples and 4 apples to get 10 apples. We should not insist gratuitously on the rigors of precise language, since that inhibits lively discussion; but we should know what we mean, and so should the people with whom we communicate.[15]

In a recent U.S. Children's Television Workshop production "Square One," comprising 75 half-hour programs for children in grades 3 through 6, an actor playing the role of a woman mathematician says "50 miles per hour times 4 hours is 200 miles." We claim that no mathematician would say any such thing; instead, she would say "50 miles per hour for 4 hours, 50 times 4 equals 200, so that's 200 miles." The maxim "speed times time equals distance" is best interpreted as a rule telling you to use the mathematical model of multiplication to obtain the number representing distance when given the numbers representing speed and time. We believe that this distinction between the real world and the model would eliminate much of the confusion surrounding the topic of rates and ratios. Mathematically, a *ratio* is just a rational number, but unfortunately rational numbers tend to remain unmentioned in the elementary curriculum, with much consequent detriment to the study of fractions; and a *rate* is a real-world quantity represented by a ratio in the model.

Let us combine the spirits of these two examples and ask the question, "If my speed now is 6 mph, and I am *accelerating* at the rate of 4 mph^2,

---

[15]This is an application of the Principle of Licensed Sloppiness (see Chapter 9 of [4]).

what will my speed be in one hour's time?" The mathematical model is again $6 + 4$, but we are certainly not forming the disjoint union of two collections. Moreover, we hardly want to keep the rather complicated units around in the mathematical model. Of course, the units have a vital role to play in the *scientific* model and in the interpretation of our mathematical result — but that's another story.

Now we will discuss the principal example of this section, our salutary paradox. We call it

The Problem of the Sailors, the Monkeys and the Coconuts

The problem is the following:[16] A group of s sailors find themselves at dusk marooned on a desert island with m monkeys and c coconuts. The sailors wish to divide the coconuts evenly among themselves, but are so tired they agree to wait until morning. However, they are very hungry sailors and each fears he won't get enough coconuts to survive. So during the night each sailor in turn wakes up, goes to the pile of coconuts, hands a coconut to each monkey (to keep them quiet), divides the remainder into s equal piles, takes his own pile, and leaves the rest. The question is — what is the smallest value of c (as a function of s and m) that makes this possible? As an example, let $s = 3$, $m = 7$. Our general solution will show that then $c = 40$; let's check it now. Given 40 coconuts, the first sailor gives 7 to the monkeys, takes 11 for himself, and leaves 22. The second sailor gives 7 to the monkeys, takes 5 for himself, and leaves 10. The third sailor gives 7 to the monkeys, takes 1 for himself, and leaves 2. Of course, when the sailors awake in the morning they will realize that the pile is smaller, but, since each one has cheated during the night, no one will say anything — perhaps they will all blame the monkeys.

We now construct a mathematical model to solve the general problem. Let u_0 be the original number of coconuts and let u_i be the number of coconuts left by the ith sailor, $i = 1, 2, \ldots, s$. It is then easy to see that

$$u_{i+1} = \frac{s-1}{s}(u_i - m), \quad i = 0, 1, \cdots, (s-1) \qquad (1)$$

We seek the smallest value of u_0 such that each u_i is a positive integer. Since obviously $u_i > 0$ if $u_{i+1} > 0$, this is equivalent to seeking the smallest value of u_0 such that each u_i is an integer and $u_s > 0$.

---

[16]Remember, the real-world is the world of the student.

We now reason within the model, *using arguments that have no meaning whatsoever in our original "real-world" problem.*[17] We ask for the ***fixed point*** of the transformation f defined by (1), that is, we consider the transformation

$$f(x) = \frac{(s-1)}{s}(x - m) \tag{2}$$

and we seek a number x such that $f(x) = x$. Simple algebra shows that the unique fixed point is $x = -(s-1)m$. Let us write u for the fixed point, so that $u = -(s-1)m$.

To the sailors — and to the critics of our approach to applying mathematics (see [2]) — this is all meaningless nonsense, "abstract," "symbol-manipulation." However, it is, in fact, crucial to the solution of our problem. For we see that if we set $v_i = u_i - u$, that is, if

$$v_i = u_i + (s-1)m \tag{3}$$

then (1) is transformed into $v_{i+1} = \frac{s-1}{s}v_i$, or

$$v_i = \frac{s}{s-1}v_{i+1} \tag{4}$$

Thus

$$v_0 = \frac{s}{s-1}v_1 = \left(\frac{s}{s-1}\right)^2 v_2 = \cdots = \left(\frac{s}{s-1}\right)^i v_i = \cdots = \left(\frac{s}{s-1}\right)^s v_s \tag{5}$$

We require that each v_i be an integer with (see (3)) $v_s > (s-1)m$. Now, (5) shows that v_0 is an integer if and only if v_s has the form $k(s-1)^s$, where k is an arbitrary integer. Then $v_0 = ks^s$ and $v_i = k(s-1)^i s^{s-i}$, so that each v_i is an integer. Finally, the condition $v_s > (s-1)m$ translates to

$$k > \frac{m}{(s-1)^{s-1}} \tag{6}$$

Since $u_0 = v_0 - (s-1)m = ks^s - (s-1)m$, we may announce the solution of the problem. The minimum value of c is $ks^s - (s-1)m$, where k is the smallest positive integer satisfying (6). Plainly,

if $s = 3, m = 7$, then $k = 2$ and $c = u_0 = 2 \times 27 - 2 \times 7 = 40$

We have already drawn attention to a key feature of our solution, namely, that the mathematical model incorporated features uncorrelated to the con-

---

[17]This is the paradox. We make essential use of negative numbers in our argument, but negative numbers of coconuts are unheard of on the island. Nor could there conceivably be a "fixed point" for the function describing the sailor's raids on the coconut supply.

text of the original problem but essential to its solution — this shows how subtle is the notion of relevance in mathematics! Notice also, in this connection, that condition (6) involves a fraction on the right-hand side, although k must be a whole number and *fractions are excluded by the context*.

However, there are other points about mathematical method illustrated by this example. First, our method, involving a study of the *linear recurrence relation* (1), placed us firmly within an important mathematical context. Thus, once the mathematical model has been identified, it is the *mathematical* context, and not the *physical* context, which should determine our strategy.

Second, we find that our mathematical model has provided us with much additional information beyond the mere answer to our original question — and *some of that information is highly relevant to our real-world situation*. For example, our solution gives us *all* possible values of c, namely, $ks^s - (s-1)m$, and not merely the smallest possible value. As another example, suppose we restrict our original problem so that $s \geq m$ (this restriction is implicit in some published versions of the problem). We then see that condition (6) is satisfied by $k = 1$, except in the trivial case $s = 2$. Thus the solution if $s \geq m$, excluding the trivial case, is

$$c = s^s - (s-1)m \tag{7}$$

We remark that if we hold s fixed in (7), then c is a *decreasing* function of m; that is, the more monkeys, the fewer coconuts needed to render the process possible — a counter-intuitive conclusion which would surely surprise the monkeys as much as it does us!

This last remark illustrates our experience that orienting a course of mathematical instruction around *problems* is, really, a difficult thing to do well. It is not sufficient to pose problems and discuss their solution; for the solution answers questions one could not reasonably have expected, in advance, even to ask. Even more, the solution will itself almost certainly suggest questions, perhaps of a purely mathematical nature, that it would be advantageous to consider even if the original problem is entirely solved; and such questions may have nothing to do with the context of the original problem. We leave our readers with the challenge to find such questions in the mathematics discussed above — and elsewhere.

● ● ● **BREAK 5**

(1) Show that exactly the same mathematical model works if the monkeys are so cantankerous that monkey j can only be placated by receiving, each time, n_j coconuts, instead of just one.

(2) Suppose that $u_{i+1} = d(u_i - m)$, $i = 0, 1, \cdots$. Express u_i as a function of u_0.

• • • BREAK 6 (The Philanthropist's Paradox[18])

An American philanthropist went up to a poor man in the street and showed him two envelopes. "I have put some money into each of these envelopes. There is twice as much in one envelope as in the other. You can take whichever envelope you like." The poor man chose one, opened it, and found $100 in it. The philanthropist then said, "If you like, you can return the $100 and take the other envelope." The poor man thought to himself, "There are equal chances that the other envelope contains $200 and $50. Thus my expectation, if I choose the other envelope, is $\frac{200+50}{2} = 125$ dollars." So he chose the other envelope.

One concludes that the correct strategy when one meets the philanthropist with the two envelopes is to choose one and then alter one's choice and choose the other. Discuss!

REFERENCES

1. Beckenbach, Edwin F., Baseball Statistics, *The Mathematics Teacher* **72**, May (1979), 351–352.
2. Friedlander, Richard J., Ol' Abner has done it again, *American Mathematical Monthly* **99**, No. 9 (1992), 845.
3. Hilton, Peter, The emphasis on applied mathematics today and its implications for the mathematics curriculum. In P. J. Hilton and G. S. Young (eds.), *New Directions in Applied Mathematics* (pp. 155–163): Springer, 1985.
4. Hilton, Peter, Derek Holton, and Jean Pedersen, *Mathematical Reflections — in a Room with Many Mirrors*, 2nd printing, Springer-Verlag NY, 1998.
5. Saari, Donald G., and Fabrice Valognes, Geometry, voting and paradoxes, *Mathematics Magazine* **71**, No. 4 (1998), 243–259.
6. Schwenk, Allen J., Beware of Geeks Bearing Grifts, *Math Horizons* (published by the Mathematical Association of America), April 2000, 10–13.

[18] We are indebted to Dr. David Fisher for acquainting us with this intriguing paradox.

7. Simpson, E. H., The interpretation of interaction in contingency tables, *J. Royal Stat. Soc.*, Ser. B, 13, No. 2 (1951), 238–241.

8. Stewart, Ian, Mathematical recreations, *Scientific American*, November (1997), 77–99.

9. Webb, John, Tournament Paradox, *Mathematical digest* **44** (1972), 21.

10. Voles, Roger, Rings of winning dice and spinners, *Mathematical Gazette*, Vol. 83, No. 497, July (1999), 298–303.

2

Not the
Last of Fermat

C H A P T E R

2.1 INTRODUCTION: FERMAT'S LAST THEOREM (FLT)

A lot has been written about Fermat's Last Theorem since its proof was announced in 1993.[1] What we write here will undoubtedly not be the last that's written on the subject.

As it took Andrew Wiles, a brilliant mathematician, seven years to put together the deepest mathematical proof of the last century, there's no way that we are going to be able to take you step by step through his proof. In fact, relatively few mathematicians could actually do that anyway, however strong their mathematical background. What we will try to do, instead, is to give you the background to the problem and an idea of the strategy adopted by those involved in solving it — and of the human perspiration expended in the final fall of the theorem. What follows then is a middle course between Singh's best-selling book [8] and the more mathematical, and more sharply focused, treatments to be found in Cox [1], Gouvêa [3], Mazur [5], and Ribenboim [7]. If you become interested in the topic of this chapter, we recommend you first read [8] and then, if you have some undergraduate mathematics under your belt, try [1], [3], [5], or [7]. For more background material there is also [9]. So you should think of this chapter as not the end but rather the beginning of what there is to know about

---

[1] The announcement was premature! There was a gap in the argument, but a complete proof did appear in 1994.

Fermat's Last Theorem. Certainly we are making no claim whatsoever that our exposition is complete.[2]

2.2 SOMETHING COMPLETELY DIFFERENT

Freda keeps some beetles and some spiders. Altogether her pets have 48 legs. How many spiders does she have?

If you play around with this problem for a while, you might suspect that it contains insufficient information. After all, if Freda has b beetles and s spiders we can only get one equation,

$$6b + 8s = 48$$

This equation holds simply because beetles have 6 legs and spiders have 8. But we have one equation that has *two* things we don't know: s and b. Normally in problems like this we need another equation. However, nobody thought to give us one. So what do we do now?

One step that might simplify things a little is to divide through by the common factor 2. Is

$$3b + 4s = 24$$

any better? Well, we want to find s, so rearrange the equation to give

$$4s = 24 - 3b$$

And then you might see that the RHS of the last equation has a factor 3:

$$4s = 3(8 - b)$$

So what? Remember that b and s are *positive integers*.

So $4s$ must also be divisible by 3. That surely means that s is divisible by 3. OK, then $s = 3k$ for some integer k. What should we do next? Perhaps we'll make some progress if we replace s by $3k$ in the equation:

$$4(3k) = 3(8 - b)$$

so

$$4k = 8 - b$$

Now it's clear that the RHS is no greater than 8, so $k = 0$, 1, or 2. Hang on! If $k = 0$, there are no spiders, and we're told that Freda has "some spiders." And if $k = 2$, then $b = 0$, but she's got "some beetles." This must mean then that $k = 1$, so $s = 3k = 3$. Freda has three spiders.

---

[2]Michael Rosen, in his favorable review of [7] (*Notices of the AMS*, **47**, No. 4 (2000), 474) wrote, "It is somewhat sad that no one expects any longer that an elementary proof of FLT will ever emerge."

● ● ●　**BREAK 1**

Try this one. At the Post Office, Dennis spent exactly $2 on stamps. He bought some 4¢ stamps, ten times as many 2¢ stamps and made up the balance with 10¢ stamps. How many stamps of each denomination did he buy?

Let's try one more. The other day José cashed a check at the bank. The teller accidentally interchanged the dollars and cents values. When later, José could only get 49¢ for his 1963 Cadillac he thought he was having a bad day. But then he realized that he now had twice as much cash as he had written the check out for. What was the real value of the check?

Doing this the traditional way would probably mean assuming that the original check was for x dollars and y cents. So the teller had given José y dollars and x cents. After selling the Cadillac he had y dollars and $(x + 49)$ cents. But this turned out to be twice the value of the original check — exactly $2x$ dollars and $2y$ cents. This gives us an "equation"

$$y \text{ dollars} + (x + 49) \text{ cents} = 2x \text{ dollars} + 2y \text{ cents}$$

Probably we won't get anywhere with this until we change it all into cents. At that point we get a genuine linear equation

$$100y + (x + 49) = 200x + 2y$$

Once again we have one equation in two unknowns; but once again we know that x and y are positive integers. How can we solve our equation? Probably in much the same way as we did with the insects. But let's tidy it up first. The equation then becomes

$$98y + 49 = 199x$$

There seems to be a factor of 49 on the left, so we rewrite the equation as

$$49(2y + 1) = 199x$$

Since 199 and 49 have no factors in common and 49 is a factor of the LHS of this equation, then x must be divisible by 49. So let $x = 49k$. We then get

$$2y + 1 = 199k$$

Did that help? What do we know about y? It can't be bigger than 99, because it was the cents amount on José's check. Now, since $y \leq 99$

$$199k = 2y + 1 \leq 198 + 1 = 199$$

So $k = 0$ or 1. If k were zero, then y would be negative, which is absurd. (The teller would surely have noticed a check made out for minus half a cent!) This means that $k = 1$ and $x = 49$. What's more, $2y + 1 = 199$, so $y = 99$, and José's check was for \$49.99.

• • • BREAK 2

Check the last answer just in case we've made a mistake.

The point of this section, is to show that not all the information you need is given explicitly in a problem or contained in the equation(s) you obtain from the data. More than that, the fact that some problems are about integers, and even positive integers, can be as helpful as having another equation. When you think about it, we couldn't have solved $6b + 8s = 48$ if spiders didn't come in whole numbers. And we would have been all at sea with José's problem if he had written out his check for non-integer values of dollars and cents (and if there were more than 100 cents to the dollar).

2.3 DIOPHANTUS

It turns out that we know rather little about the Greek mathematician Diophantus. We certainly don't know where he was born or when he was born. We might know how old he was when he died. Supposedly, the following was carved on his tomb.

> God granted him to be a boy for the sixth part of his life, and adding a twelfth part to this, He clothed his cheeks with down; He lit him the light of wedlock after a seventh part, and five years after his marriage He granted him a son. Alas! Late-born wretched child; after attaining the measure of half his father's full life, chill Fate took him. After consoling his grief by this science of numbers for four years he ended his life.

As we were saying then, he lived to a ripe old age for someone who flourished about 250 A.D. (but he may have been living as early as 150 A.D. or as late as 360 A.D.). His fame comes from the six extant books of the thirteen-volume set he wrote called *Arithmetica*. This was a treatise that he compiled while he was in Alexandria, the center of intellectual life in the Mediterranean from about 350 B.C. until about 640 A.D. Diophantus' *Arithmetica* was to number theory what Euclid's *Elements* was to geometry. While he was in Alexandria, Diophantus collected and invented a range of number-theoretical problems. He was particularly interested in problems whose solutions were rationals. However, problems involving integers are now known as ***Diophantine problems***. Similarly equations whose solutions are required to be whole numbers are called ***Diophantine equations***.

So the two problems of Section 1 (three if you include the one in BREAK 1) are Diophantine problems. The equations

$$6b + 8s = 100 \quad \text{and} \quad 98y + 49 = 199x$$

are Diophantine equations.

The library at Alexandria, which held hundreds of thousands of books containing all the knowledge that had been accumulated by the Greeks, was finally destroyed. The destruction began at the hands of Christians in 389 A.D. They were out to destroy all pagan monuments, and the library was housed in what was once an Egyptian temple. What books survived the Christians were largely destroyed by Moslems in 642 A.D. So it may be somewhat surprising that even six volumes of Diophantus' magnum opus survived.

2.4   ENTER PIERRE DE FERMAT

It's not clear that you would necessarily have gotten along very well with Pierre de Fermat. If you were an English mathematician, it's almost certain that you wouldn't! Pierre de Fermat was born into a rich family in the southwest of France in 1601. His family's fortune allowed him a good education, and he entered the civil service. His job was to make a preliminary assessment of petitions to the King of France. If someone wanted to petition the king, they had to go through Fermat. If he wasn't convinced of the merit of the case, then the petition went no further. Because of his role in society, Fermat had few social contacts. It was felt that, if a petition was put forward by one of his associates, Fermat (or indeed any other of the councillors at the Chamber of Petitions) might be swayed by his friendship to support it. Consequently, Fermat led a very solitary life. But this did give him the opportunity to engage in his lifelong interest of mathematics. In fact, he was so good at mathematics that he has been called the Prince of Amateur Mathematicians.

Today, most mathematicians are eager to publish their research results in journals that are readily available to their peers. In seventeenth-century Europe, though, mathematicians were more secretive. There was a tendency to throw out problems as a challenge. Mark you, these were problems that the poser already knew how to solve.

Fermat engaged in this practice and was particularly keen to embarrass the English mathematicians. At one stage he discovered that a particular square and a particular cube had only one number between them. What's more, he found that this was the only instance where a square and a cube differed by 2. Fermat gained particular satisfaction from the fact that the English mathematicians had to admit that they were unable to solve his square and cube problem.

• • • BREAK 3

Can *you* find one solution in integers of the equation $y^3 = x^2 + 2$?

It's worth remarking here that in 1918 the English mathematician L.J. Mordell was able to show that there are only a finite number of integer solutions to

$$y^3 = x^2 + k$$

for *any* integer k, positive or negative. But perhaps a lag of 280 years or so is somewhat excessive for the English to claim a victory, even if k is much more general than 2.

Somehow, Fermat came across a copy of Diophantus' *Arithmetica*. What he had was a translation into Latin, which, for some fortunate but unknown reason, had particularly large margins. As Fermat went through the book he frequently made comments in those margins. These may well have been lost to posterity had it not been for the fact that his son, Clément-Samuel, put together all Fermat's notes, letters, and marginal jottings and published them in a special edition of the *Arithmetica*. This volume contained many challenging problems. Eventually, only one of these remained unsolved. It was the final challenge. Because Fermat said that he had a proof, it was called Fermat's Last Theorem. However, given the fact that it remained unsolved for over 350 years, there's a very good chance that Fermat did not have a complete proof of the "theorem." Really we should have called it Fermat's Last Conjecture. But it's too late to rename it now. So what is Fermat's Last Theorem all about?

2.5 FLASHBACK TO PYTHAGORAS

Pythagoras is well known for several things. In fact we know rather more about him than we do about Diophantus. Pythagoras lived in the sixth century B.C. and wandered around the Eastern Mediterranean area until he set up his scholastic community in southern Italy. That community is known for realizing the links between harmonics and the ratios of lengths in a lyre string. They are also supposed to have sacrificed 100 oxen when Pythagoras' theorem was proved. But no one knows who actually proved the result. Anything that Pythagoras and his followers discovered was considered the property of the community.

Pythagoras' Theorem   *In a right-angled triangle, the square on the hypotenuse is equal to the sum of the squares on the other two sides.*

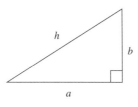

That is, $h^2 = a^2 + b^2$.

One important thing to note here is that Pythagoras did actually prove that this result was true *for all right-angled triangles*. A thousand years before Pythagoras, the Chinese and the Babylonians knew specific values of a, b, and h — for specific right-angled triangles. Pythagoras' contribution was proving the theorem for *any* right-angled triangle.

● ● ●   **BREAK 4**

(1) Give a proof of Pythagoras' Theorem.
(2) Give another proof.

Pythagoras had a lot in common with Diophantus. Pythagoras very much preferred dealing with integers and fractions.

Pythagoras and his disciples were able to find an infinite number of right-angled triangles whose sides had integer lengths. You may remember a few of the following:

$$3^2 + 4^2 = 5^2$$
$$5^2 + 12^2 = 13^2$$
$$7^2 + 24^2 = 25^2$$
$$9^2 + 40^2 = 41^2$$

You may even be able to see a pattern here.

Three integers that can be the sides of a right-angled triangle are known as **Pythagorean triples** and are sometimes represented by the notation (a, b, h). We've seen that $(3, 4, 5)$, $(5, 12, 13)$, and so on, are Pythagorean triples. Now, it turns out that $(u^2 - v^2, 2uv, u^2 + v^2)$, for u, v integers with $u > v$, are Pythagorean triples. So, by using different values of u and v, we can get a number of Pythagorean triples. In fact, we can get an infinite number, as we shall soon see.

● ● ● BREAK 5

Show that $(u^2 - v^2, 2uv, u^2 + v^2)$ is indeed a Pythagorean triple.

Before we get started on the following arguments, let's agree to use the abbreviation PT for Pythagorean triple. Now, we will say that a PT (a, b, h) is *primitive* if a, b, h are pairwise coprime. We now prove two key results.

Theorem 1   *Every PT is a multiple of a primitive PT. Moreover, if*

$$(a, b, h) = \lambda(a_1, b_1, h_1)$$

with (a_1, b_1, h_1) primitive, then

$$\lambda = \gcd(a, b) = \gcd(a, h) = \gcd(b, h) = \gcd(a, b, h)$$

Thus (a_1, b_1, h_1) is primitive if and only if $\gcd(a_1, b_1, h_1) = 1$.

Proof   Let $\lambda = \gcd(a, b)$. Then $\lambda | a$, $\lambda | b$, so $\lambda^2 | a^2$, $\lambda^2 | b^2$, and thus $\lambda^2 | h^2$. But if $\lambda^2 | h^2$, then certainly $\lambda | h$. It follows that

$$\gcd(a, b) = \gcd(a, b, h)$$

Similarly, we can prove that

$$\gcd(a, h) = \gcd(a, b, h)$$

and

$$\gcd(b, h) = \gcd(a, b, h) = \lambda,$$

say. Set

$$a = \lambda a_1, \quad b = \lambda b_1, \quad h = \lambda h_1$$

Then $\gcd(a_1, b_1, h_1) = 1$, so (a_1, b_1, h_1) is certainly a primitive PT. Moreover, if (a, b, h) is primitive, then

$$\gcd(a, b) = \gcd(a, h) = \gcd(b, h) = 1,$$

so $\gcd(a, b, h) = 1$. This completes the proof.   □

Now let (a, b, h) be a primitive PT. We claim that one of a, b is odd and the other even, and that h is odd. For we certainly can't have a, b, h all even, since (a, b, h) is primitive; and we can't have a, b both odd, since $a^2 \equiv 1 \pmod 4$ and $b^2 \equiv 1 \pmod 4$ implies $h^2 \equiv 2 \pmod 4$, which is clearly impossible. Suppose, then, that a is odd, b is even, and h is odd.

Theorem 2   *Let (a, b, h) be a primitive PT with a odd, b even, h odd. There are then unique integers u, v such that $u > v$ and*

$$a = u^2 - v^2, \quad b = 2uv, \quad h = u^2 + v^2$$

moreover, u, v are coprime and of opposite parity.[3]

Proof   We have $a^2 + b^2 = h^2$, so

$$\left(\frac{b}{2}\right)^2 = \left(\frac{h+a}{2}\right)\left(\frac{h-a}{2}\right) \tag{1}$$

but $\frac{h+a}{2}$, $\frac{h-a}{2}$ are coprime, for any factor of $\frac{h+a}{2}$ and $\frac{h-a}{2}$ is a factor of h and of a (why?). But (a, b, h) is primitive, so $\gcd(a, h) = 1$. Thus it follows from (1) that each of $\frac{h+a}{2}$, $\frac{h-a}{2}$ is a perfect square, say

$$\frac{h+a}{2} = u^2, \quad \frac{h-a}{2} = v^2,$$

u, v coprime. But then $h = u^2 + v^2, a = u^2 - v^2$, and $(\frac{b}{2})^2 = u^2 v^2$, so $b = 2uv$. Since a is odd, u and v must have opposite parity. Finally, we prove uniqueness. For if we also have $h = u_1^2 + v_1^2, a = u_1^2 - v_1^2$, $b = 2u_1 v_1$, then

$$u^2 + v^2 = u_1^2 + v_1^2, \quad u^2 - v^2 = u_1^2 - v_1^2,$$

showing that $u^2 = u_1^2, v^2 = v_1^2$, so $u = u_1, v = v_1$.   □

Notice that we can extend the uniqueness statement to arbitrary PTs. Thus, if (a, b, h) is a PT, it is uniquely expressible as

$$(a, b, h) = \lambda(a_1, b_1, h_1)$$

where (a_1, b_1, h_1) is primitive, since λ is determined as $\gcd(a, b, h)$; hence (a, b, h) is uniquely expressible as

$$\lambda(u^2 - v^2, 2uv, u^2 + v^2) \tag{2}$$

where u, v are coprime of opposite parity, and we assume a_1 odd.

---

[3]This means that one is odd and the other is even.

• • •   BREAK 6

(1) Which of the following are primitive PTs and which are not? For those that are, find the values of u and v that yield the given PT.

$$(3, 4, 5); \quad (5, 12, 13); \quad (231, 108, 255); \quad (99, 20, 101).$$

(2) Write those that are not primitive PTs as a product $\lambda(a, b, h)$ as in Theorem 1, and then find the corresponding u and v.

2.6   SCRIBBLES IN MARGINS

Apparently, Fermat was working on Diophantus' *Arithmetica* in about 1637. He was fascinated by Pythagoras' Theorem and Pythagorean triples. While working in this area of Diophantus' book, Fermat was inspired to write (in Latin)

> To divide a cube into two cubes, a fourth power into two fourth powers, and in general any power above the second into two powers of the same denomination, is impossible. Of this I have assuredly found a marvelous proof, but this margin is too narrow to contain it.

What Fermat had first done was to do what mathematicians continually try to do. He had attempted to generalize. Pythagorean triples are positive-integer solutions to the equation

$$x^2 + y^2 = z^2.$$

Fermat had asked the obvious mathematical question: Do the Diophantine equations

$$x^3 + y^3 = z^3,$$
$$x^4 + y^4 = z^4,$$
$$x^5 + y^5 = z^5,$$

etc. have any positive-integer solutions? He had worked on them and decided that they did not. He then thought that he could prove it.

Fermat's Last Theorem   *For $n \geq 3$ an integer, there are no positive-integer solutions of*

$$x^n + y^n = z^n$$

But did he really have a proof? We think the answer has to be "most likely not." There are at least two compelling reasons for this. The first is that Fermat had made other claims of this nature that were doubtful. For instance, the problem of the cube and square that differ by 2, which Fermat used to torture the English, appears not to have been satisfactorily solved by Fermat. (This may be little consolation to the poor English mathematicians of the seventeenth century.)

The second reason is that, whatever proof he thought he had, it would have been elementary from a modern standpoint. Consequently, it would be extremely surprising that no one in the last 300 years had been able to find Fermat's proof.

What seems to be quite possible is that Fermat made a mistake. He thought he had a proof, but there was an error. Mathematicians do make errors. As you will see in Chapter 5, Kempe made an error when he thought he had proved the Four Color Theorem. As you will see in Section 9 of this chapter, Wiles, who ultimately put the final nail in the coffin of Fermat's Last Theorem, made an error in his first "proof" which was made public in 1993.

It is unlikely, however, that Fermat was attempting to mislead. There is no evidence of his doing anything like that on any other occasion. And he seems to have mentioned only the cases $n = 3$ and $n = 4$ in correspondence with other mathematicians. It is perhaps likely that he meant to delete his marginal comment but forgot to do so. What's more, if he had a conjecture rather than a theorem he would probably have admitted it. For instance, he stated once that he thought that $2^{2^n} + 1$ was always a prime number for every positive integer n. But he added that while he was pretty sure that this was the case, he didn't know how to prove it. (See Chapter 4 for a counterexample.)

On balance, we probably have to believe in Fermat's integrity. It is reasonable to believe that he thought he had a proof of FLT but that somehow he had made an error and his proof was false.

2.7   *n* = 4

The case $n = 3$, that is, the fact that $x^3 + y^3 = z^3$ has no integer solutions, was known to Arabian mathematicians some 700 years before Fermat scribbled his cryptic margin message. However, they were unable to prove it to the rigorous level required by present day mathematics. Fermat thought he had a proof, but it is not likely that he did. Surprisingly, then, the value $n = 4$ was the first case of Fermat's Last Theorem to be solved, though perhaps "solved" is not quite the right word. Fermat himself put down enough details in 1640 that we can reconstruct a complete proof. This proof was not complete by today's standards. However, it did introduce a nice new method. The cunning idea that Fermat used for his "proof" of the case $n = 4$ (and which was completed in detail by Leibniz in 1678) was the ***method of infinite descent***. The concept here is to assume that you have a solution of the equation and show that you can produce a smaller solution. Hence you can keep getting smaller and smaller solutions. This is obviously not possible when you are dealing with positive integers, so the method of infinite descent will show that there are no solutions. Another way of thinking of this is to suppose that you have the *smallest* solution. By showing a smaller solution you have a contradiction.

Now we can consider the equation $x^4 + y^4 = z^4$. Just to be perverse, though, we will apply the method of infinite descent to $x^4 + y^4 = z^2$. Let's think about this for a moment. How does the logic go? Suppose $x^4 + y^4 = z^4$ has a solution in integers $x = p$, $y = q$, $z = r$. Then $p^4 + q^4 = r^4 = (r^2)^2$. So a solution to $x^4 + y^4 = z^4$ will give a solution to $x^4 + y^4 = z^2$ (via $p^4 + q^4 = (r^2)^2$). Logically, then, if we can show that $x^4 + y^4 = z^2$ has no solutions in positive integers, then the last sentence tells us that neither does $x^4 + y^4 = z^4$.

If we then start with $x^4 + y^4 = z^2$, we can immediately relate this to Pythagoras. So $(x^2)^2 + (y^2)^2 = z^2$. We now know that we can concentrate on primitive solutions and hence suppose that

$$x^2 = u^2 - v^2, \quad y^2 = 2uv, \quad \text{and} \quad z = u^2 + v^2,$$

where u and v have no common factor and they are not both odd. Suppose now that u is even and v is odd. Then $u^2 = 4s$ and $v^2 = 4t + 1$ for some s and t. Hence $x^2 = 4(s - t) - 1 = 4(s - t - 1) + 3$. But squares can only have remainders of 0 or 1 when divided by 4. Hence u is odd and v is even. We now have

$$x^2 + v^2 = u^2$$

We then have, since (x, v, u) must be a primitive PT (remember that $\gcd(u, v) = 1$),

$$x = p^2 - q^2, \quad v = 2pq, \quad \text{and} \quad u = p^2 + q^2$$

where p, q have no factors in common and p, q are not both odd. Recall that $y^2 = 2uv$. So now we have

$$y^2 = 4pq(p^2 + q^2)$$

But p and q have no common factors, so neither do p and $p^2 + q^2$, or q and $p^2 + q^2$. This means that p, q, $p^2 + q^2$ are all squares. Hence

$$p = r^2, \quad q = s^2, \quad p^2 + q^2 = t^2, \quad \text{and} \quad \gcd(r, s) = 1.$$

Now here's the coup de grâce! This shows that

$$(r^2)^2 + (s^2)^2 = t^2$$

or

$$r^4 + s^4 = t^2$$

Now, we can measure the "size" of a primitive solution of $x^4 + y^4 = z^2$ by the size of z. If we could show that $t < z$, we'd have a primitive solution r, s, t smaller than the original x, y, z. So how is t related to z?

Now

$$z = u^2 + v^2 > u^2 = (p^2 + q^2)^2 = t^4.$$

Then $z > t^4$. So $t < \sqrt[4]{z} < z$, since $z > 1$, obviously. A solution x, y, z of $x^4 + y^4 = z^2$ has led us to a smaller solution r, s, t. Infinite descent comes into play. So Fermat's Last Theorem is true for $n = 4$.

• • • BREAK 7

Why is it that squares have only remainders of 0 and 1 on division by 4?

It's worth thinking about where we are right now. It may seem that we have only covered one of an infinite number of cases of Fermat's Last Theorem, the case $n = 4$. In fact, though, we have settled an infinite number of cases. Why is this so? Have a look at $x^8 + y^8 = z^8$. Suppose r, s, t is a solution to this. Then

$$r^8 + s^8 = t^8$$

and

$$\left(r^2\right)^4 + \left(s^2\right)^4 = \left(t^2\right)^4$$

A solution to $x^8 + y^8 = z^8$ would then give us a solution to $x^4 + y^4 = z^4$. We have just shown that the last equation has no integer solutions. So $x^8 + y^8 = z^8$ has no integer solutions. It's the same little trick that we used in going from knowing that $x^4 + y^4 = z^2$ has no solutions to knowing that $x^4 + y^4 = z^4$ has no solutions.

As a result of all that, then, we are now able to assert confidently that $x^{12} + y^{12} = z^{12}$ has no integer solutions, that $x^{16} + y^{16} = z^{16}$ has no integer solutions, that .... Fine, so we now know that, for any positive integer m, $x^{4m} + y^{4m} = z^{4m}$ has no integer solutions.

This same argument now puts us in a position where we only have to show that Fermat's Last Theorem is true when n is an odd prime. If we could prove that $x^p + y^p = z^p$ has no integer solutions for all odd primes p, then the truth of Fermat's Last Theorem for any composite number n would follow. This is because, if $n \geq 3$, then either n is a power of 2 divisible by 4 or $n = pq$ for some odd prime p. Since $x^4 + y^4 = z^4$ has no integer solutions, we have already settled the first case. On the other hand, if $n = pq$, then $x^{pq} + y^{pq} = z^{pq}$ gives $(x^q)^p + (y^q)^p = (z^q)^p$. The usual argument now reduces the cases we have to consider for n. We thus have to settle FLT only for n an odd prime.

2.8   EULER ENTERS THE FRAY

Mathematics is a hard taskmaster. We have seen how excited mathematicians get when they prove a theorem — remember the story of Pythagoras and the 100 oxen. We have noted the incompleteness of Fermat's attempts at producing proofs. But mathematics has standards. The full proof must be foolproof.

Although Fermat thought that he had settled the case $n = 3$ by the same infinite descent approach that he had used with the case $n = 4$, there is no evidence of this. In fact, it is generally accepted that the first person to prove FLT for $n = 3$ was the Swiss mathematician Leonhard Euler.

Euler is known for his contributions to many areas of mathematics. You will see his work on the Königsberg bridges problem in Section 5.4. Then there is the result that $a^{\phi(n)} \equiv 1 \pmod{n}$, where ϕ is Euler's totient function[4] (see Chapter 2 of [4]). And there is his polyhedral formula $V - E + F = 2$, which links the number of vertices, edges, and faces of a polyhedron homeomorphic to a sphere. Euler did much, much more. He was extremely prolific.

In 1753, Euler was able to settle the case $n = 3$ of Fermat's Last Theorem. This was the first development in Fermat's Last Theorem in over 100 years. However, there was a step in the proof that happened to be correct for the situation where Euler applied it, but it was a step that would not work in general. Strictly speaking, Euler should have justified this step. Now the general approach to Fermat's Last Theorem seemed to be to try to knock off a case at a time. This was a bit like proving that any map with

---

[4] In fact, $\phi(n)$ is the number of positive integers m such that $m \leq n$ and m is prime to n.

10 countries is 4-colorable, and then that a map with 11 is 4-colorable, and so on (see Chapter 5). While everyone knew that this case-by-case approach was never going to solve either problem, there was always the hope that, after enough cases had been dealt with, someone would see a general approach or find a counterexample.

One person who tried to move away from this case-by-case strategy was one of the outstanding pre-twentieth-century women mathematicians, Sophie Germain. Her attack lay in looking at a special class P of primes, namely, those primes p for which $2p + 1$ is also a prime. This class obviously includes 5 and 11 but not 7 or 13. For primes of this class, Germain was able to show that if there is a solution of $x^p + y^p = z^p$, then one of x, y, or z would be a multiple of p. This result could then be used to restrict the set of possible solutions.

As a result of the high interest in the problem generated partly by Germain's new idea, the French Academy established a series of prizes for a solution to Fermat's problem. This was not to be the last such prize.

In 1825, as a result of Germain's work, two famous mathematicians, Peter Gustav Lejeune Dirichlet and Adrien-Marie Legendre, working independently, managed to settle the case $n = 5$.

But still the method used by Fermat and Euler was popular. Gabriel Lamé thought that he would be able to use this approach to solve Fermat's Last Theorem not just for some specific cases, but for all odd primes. This was in 1847. And it was the German mathematician Ernst Kummer who first was able to point out that the approach would not work in general, and second was able to show that it was valid for a large class of primes called the regular primes (for a definition of a regular prime, see [1]).

What Lamé (and Euler) had done was to make a very simple oversight. He had assumed that, in the realms in which he had been working, unique factorization occurred. This is a perfectly natural assumption. In ordinary arithmetic this always holds. For instance, $1050 = 2 \times 3 \times 5^2 \times 7$, and this is the only way to factorize 1050 expressing the result as a product of primes (irreducibles). Taking a simpler example, $6 = 2 \times 3$, the factorization is unique. It can only be done in one way among the positive integers, assuming that one ignores the order of the factors.

But the way that the proofs had been developing required calculations to take place among the complex numbers. A complex number is one of the form $a + bi$, where $i = \sqrt{-1}$. For instance, $3 + 4i$ and $2 - 3i$ are two complex numbers. If you want to do arithmetic with them you just treat the i as an algebraic quantity except that, if you ever get i^2, you convert it to -1. So

$$(3 + 4i) + (2 - 3i) = 5 + i$$
$$(3 + 4i) - (2 - 3i) = 1 + 7i$$

and

$$(3 + 4i)(2 - 3i) = 6 + 8i - 9i - 12i^2$$
$$= 6 - i - 12(-1)$$
$$= 6 - i + 12$$
$$= 18 - i$$

Furthermore,

$$\left(1 + \sqrt{5}i\right)\left(1 - \sqrt{5}i\right) = 1 - 5i^2 = 6$$

Ah! $(1+\sqrt{5}i)(1-\sqrt{5}i) = 2 \times 3$, where each of the four (complex) numbers is irreducible in the set of numbers $a + b\sqrt{5}i$, with a, b ordinary integers. So, using the complex numbers that were relevant for the Euler–Fermat proof of FLT, it was possible for numbers to be factorized in more than one way. It is possible for 6 to be factorized as 2×3 or as $(1 + \sqrt{5}i)(1 - \sqrt{5}i)$.

This problem of unique factorization had actually been a problem for Fermat, though he hadn't realized it. Let's go back to the equation $x^3 = y^2 + 2$, which Fermat claimed had a unique solution in positive integers. His method of proving this was to note that

$$x^3 = y^2 + 2 = \left(y + \sqrt{2}i\right)\left(y - \sqrt{2}i\right)$$

Now, since the left-hand side of the equation is a cube and since the two factors on the right have no factors in common, then, given unique factorization, these factors must both be cubes. Hence

$$y + \sqrt{2}i = \left(p + q\sqrt{2}i\right)^3 = p^3 + 3p^2q\sqrt{2}i + 6pq^2i^2 + 2q^3\sqrt{2}i^3$$
$$= p^3 - 6pq^2 + \sqrt{2}\left(3p^2q - 2q^3\right)i$$

where p and q are integers. This implies that $3p^2q - 2q^3 = 1$, and so $q(3p^2 - 2q) = 1$. Since p and q are integers, $q = 1$ or -1. Hence, if $q = 1$, $3p^2 - 2 = 1$. So $p = \pm 1$. This means that $y = p^3 - 6pq^2 = \pm 5$. Since y is a positive integer, $y = 5$, and hence $x = 3$.

• • • **BREAK 8**

What happens if $q = -1$ in the above argument?

The possible lack of unique factorization was something that Fermat overlooked in his $x^3 = y^2 + 2$ problem. (Could it have been the reason that the English mathematicians couldn't do it?) However, it was also an insurmountable difficulty for a certain class of odd primes, in one approach to FLT. For the rest of the primes, the regular primes, Kummer was able to show that the Euler–Fermat style of proof would go through. But how to tackle the irregular primes?

Apparently, Lamé was devastated when Kummer's letter about the regular and irregular primes was read to the French Academy of Sciences in 1847. It was no consolation that Lamé had proved the case $n = 7$ in 1839. He had hoped that he could get much further.

Again there was a long period of little development. The next significant progress occurred in 1908 when Dickson proved that FLT was true for all n up to 7000.

In 1908, Paul Wolfskehl died and left 100,000 marks for anyone who could prove the truth of FLT. There is a fascinating story about this. Apparently, Wolfskehl had been spurned in love and decided to commit suicide. He nominated a time and place for himself and put everything in readiness for the final event. Having got everything nicely organized ahead of time, he had some time to kill before he pulled the trigger. Being an amateur mathematician of some note, he started to read a book on number theory. He became interested in Kummer's proof for regular primes and was surprised to find that Kummer had made a mistake. Wolfskehl tried to patch up Kummer's proof and became so engrossed in the problem that the appointed time of his suicide passed unnoticed. As it happened, he was able to rectify Kummer's proof. Elated by this, he decided that life was worth living after all. So he gave up the idea of suicide and changed his will to include the very substantial prize of 100,000 marks for the first person to prove FLT.

Unfortunately, the prize did not have the reaction he had hoped for. Mathematicians generally seemed to have decided that FLT was too hard or not sufficiently significant to bother with. On the other hand, the prize, worth just under 2 million US dollars in today's money, attracted a great deal of interest outside professional mathematical circles. The committee responsible for overseeing the prize in the German town of Göttingen was inundated with entries. The prize was not awarded (until recently).

2.9   I HAD TO SOLVE IT

In 1963, the ten-year-old Andrew Wiles wandered into a library in Cambridge, England, and found a copy of E.T. Bell's *The Last Problem*. This presented the long history of FLT including its Greek roots. Wiles was fascinated by what he read. In [7] he says

> It looked so simple, and yet all the great mathematicians in history couldn't solve it. Here was a problem that I, a ten-year-old, could understand and I knew from that moment that I would never let it go. I had to solve it.

How many other ten-year-olds have said they had to solve it? How many other people had said they had to solve it? Possibly thousands. But none had succeeded and at that time, not much seemed to be happening on the FLT front. With the advent of computers after the Second World War, it soon became known that the theorem was true for progressively larger values of n. By 1993 it had been shown to be true for n up to 4,000,000. However, proving the theorem by computer (in general, using the computer to prove anything in mathematics) is not just a matter of starting out with a particular value of n, and checking all possible values of x, y, and z. That approach is doomed to failure from the start.[5] There are, after all, an infinite number of numbers to try for x, y, and z, so the process is, in principle, endless. So there have to be ways found to reduce the problem to a finite search. This means that some relevant mathematical arguments have to be developed and applied. Tricks like that produced by Sophie Germain are needed to make the problem a finite one.

There is, of course, another problem with the computer. It may perhaps help you to show that a particular value of n satisfies FLT, but it won't enable you to solve the entire problem. At the end of the computations you will just know that the theorem is true for another value of n. There will still remain an infinite number of numbers n to be checked. In fact, it may be better to solve for a particular value of n by hand. This way you may get some insight into the problem that you would not get by using the computer. On the other hand, to solve for a particular n by computer requires various reduction techniques. These techniques might be developed into a more general argument. But these are the hands-on parts of the process and in some sense are independent of the computer.

So there is still the need for mathematicians to invent proofs. Of course, this is the position at the end of the twentieth century. It may be that, in the future, machines will have been developed to the stage where they can

---

[5]Unless FLT turned out to be false.

find proofs of theorems. Penrose in [6] suggests that this is unlikely, but who knows for sure what the future will bring?

Anyway, while we have been chatting, Wiles has achieved his Bachelor's degree and become a graduate student under John Coates at Cambridge. In order to get a Ph.D. (Doctor of Philosophy degree) at Cambridge, and most other universities operating under the British system, it's necessary to work on a problem for 3 years or so and then submit your results in the form of a book called a thesis. The thesis is examined by a small group of people. They are looking for original material; if they don't find it, the candidate is unlikely to pass and be granted the Ph.D.

Coates decided that Wiles should work on **elliptic curves**. These are equations of the form $y^2 = Ax^3 + Bx^2 + Cx + D$, where A, B, C, D are whole numbers. We've already seen one of these in Section 3, namely, $y^2 = x^3 - 2$. You remember that Fermat claimed to be able to show that the only solution in positive integers of this equation is $x = 3$, $y = 5$.

While it may seem that Wiles was putting his ambition on hold, it is important to have a Ph.D. to be an academic mathematician; and Wiles wanted a career in academia. As it turned out, elliptic curves are fundamental to his proof of FLT. But there is another concept fundamental to his proof — that of modular functions.

★   We're afraid that from here we are going to have to skim over many of the details of Wiles' proof. In that vein we give an approximation to the definition of modular functions but refer the reader who would like something more precise to [1] or [3]. Approximately, a *modular function of level N* is a complex function f such that

$$f\left(\frac{az+b}{cz+d}\right) = f(z)$$

for all integers a, b, c, d with $ad - bc = 1$ and $N|c$. An example of a modular function is

$$16e^{\pi i z} \prod_{n=1}^{\infty} \left(\frac{1 + e^{2\pi i n z}}{1 + e^{2\pi i (n-1)z}}\right)^8$$

In the 1950s, Yutaka Taniyama and Goro Shimura began to be interested in these relatively complicated functions of complex variables. It is easy to define complex functions. For instance, $f(z) = z^2$ is an example. This means that $f(i) = i^2 = -1$,

$$f(1 + 2i) = (1 + 2i)^2 = 1 + 4i + 4i^2 = -3 + 4i,$$

and so on.

Taniyama and Shimura began to see a link between elliptic curves and modular functions. In every case they tried, they could get two modular functions to fit into a given elliptic curve. So they made the following conjecture.

The Shimura–Taniyama Conjecture: *Given an elliptic curve*

$$y^2 = Ax^3 + Bx^2 + Cx + D$$

over $\mathbb{Q}$ (the rationals) there exist non-constant modular functions $f(z)$ and $g(z)$, of the same level N, such that $f(z)^2 = Ag(z)^3 + Bg(z)^2 + Cg(z) + D$.

(For reasons we won't go into here, André Weil's name is sometimes also attached to this conjecture, and Taniyama's name is sometimes written first.)

For experts in the field, this was an astounding and exciting conjecture. It is rare that two such apparently disparate areas of mathematics as the theory of elliptic curves and the theory of modular functions turn out to be so closely linked. If the conjecture were true, it would undoubtedly mean advances in both areas. These would come because results in one area could be applied to the other.

Rather tragically, Taniyama committed suicide before his work with Shimura could be advanced very far.

The Shimura–Taniyama Conjecture attracted a considerable amount of attention. Many mathematicians were confident that it was true and even wrote papers with results worded "Assuming the Shimura–Taniyama conjecture, such and such follows." This is not an unusual thing to happen in mathematics. At one stage there was a conjecture about something called the Classification of Finite Simple Groups. Many group theorists wrote papers based on the assumption that the classification was correct. What is the value of such work? Well, first, if the conjecture turns out to be true, then we know that all the results that had assumed it are also true, so the field has progressed. And, secondly, the results based on the conjecture may lead to a contradiction. In that case the conjecture would be shown to be false. So there are potential gains both ways.

Now there is one final type of function to be introduced, whose significance is explained below. It is called a ***cusp form of weight 2 and level N***. Such a function F satisfies the identity

$$F\left(\frac{az+b}{cz+d}\right) = (cz+d)^2 F(z)$$

where a, b, c, d, N are as in the definition of a modular function, but we will not attempt to define it fully here.

The next development was somewhat surprising. Gerhard Frey decided to assume that FLT was *false*, in order to produce a contradiction. Where would that lead and what had that got to do with the Shimura–Taniyama Conjecture?

In 1982 Frey assumed that u, v, w satisfy $x^p + y^p = z^p$, where p is an odd prime. In other words, $u^p + v^p = w^p$. From this equation he produced the elliptic curve

$$y^2 = x\,(x - u^p)\,(x + v^p)$$

But this was a strange elliptic curve, so strange that it probably wasn't related to a modular form. If that was so, it would contradict the Shimura–Taniyama Conjecture; you see, Frey was feeling his way towards a proof of FLT by contradiction.

So Frey went about setting up a proof that the Shimura–Taniyama Conjecture implied FLT. He recognized that a complete proof would require the expertise of a specialist in the arithmetic of modular forms; he thought such an expert would find it fairly easy to design a proof, and was surprised that it turned out to be more difficult than he had supposed. However, Jean-Pierre Serre succeeded in showing that "Shimura–Taniyama implies FLT" would follow from a certain "level-lowering" conjecture of his own, which experts could indeed start working on.

So at this stage it required two conjectures to prove FLT. This was reduced to one conjecture when Kenneth Ribet proved Serre's level conjecture in 1986. An outline of the proof went like this. Assume that the equation $x^p + y^p = z^p$ has a solution. Then produce the Frey elliptic curve. If the Shimura–Taniyama Conjecture is true, there exists a cusp form of weight 2 and level N, suitably related to the Frey elliptic curve. By the Serre level conjecture, this implies the existence of a cusp form of weight 2 and level smaller than N. Repeating this step eventually leads to the existence of a non-zero cusp form of weight 2 and level 2. It is known that the vector space of such cusp forms is the zero space, so we have a contradiction. Hence the original assumption is false and FLT is true.

So the only hurdle left to be overcome was finding a proof of the Shimura–Taniyama Conjecture.

You can imagine that the announcement of Frey's elliptic curve, and the consequent arguments linking Shimura–Taniyama and Fermat, caused a great deal of excitement in the mathematical community. However, surprisingly few mathematicians were prepared to make the attempt to prove the Shimura–Taniyama conjecture. They felt that it was an impossible task and turned their attention to other things.

But this was the spur that Wiles needed. He had long since finished his Ph.D. and had moved to a job in the Mathematics Department at Princeton University. Now he devoted himself almost entirely to settling the Shimura–Taniyama Conjecture. He didn't expect that it would be easy. He thought it might take 10 years or so. But he decided to concentrate all his research efforts into solving this one problem.

As part of his strategy he locked himself away from his colleagues and worked single-mindedly on his own. This was actually an unusual strategy. Nowadays mathematicians prefer to meet with their peers at regular intervals to discuss their work and the work of others. They also regularly go to conferences to learn what the latest techniques and results in their area are. But Wiles, very naturally, wanted the scalp of FLT for himself, and he also did not want to be distracted by others. So working on his own in his attic, he started to learn all there was to know about the subject he was about to tackle.

In 1988, Yoichi Miyaoka described a proof of FLT to a mathematics seminar in Bonn. Wiles must surely have been very upset by this news. The "proof," however, lasted only a short period before an unpatchable difficulty was discovered.

Then, in 1993, with a great sense of theater, Wiles announced his proof at a meeting in the Isaac Newton Institute in Cambridge, England. As with Miyaoka, the news hit the front pages of the world's leading newspapers! But, once again, an error was found. Wiles tried to find a way round his mistake, but there seemed to be no way that he could patch up his original argument. It was back to the attic.

Another blow came when rumours circulated that a counterexample to FLT had been found. There was, it was claimed, some n, suitably enormous, for which integer values of x, y, z could be found such that $x^n + y^n = z^n$. An advantage of e-mail is that news like this sweeps the mathematical world very quickly. Another advantage is that the original e-mail is dated. The message that started the excitement was sent on April 1!

At this stage Wiles was not working alone. A former research student of his, Richard Taylor, came and worked with him in an effort to provide mathematical support in areas where Wiles felt he would like to have some expert cooperation. It took some time to work around the error of the 1993 proof, but Wiles and Taylor came up with a proof of a restricted form of the Shimura–Taniyama Conjecture in late 1994.[6] This time the referees who pored over the two submitted manuscripts could find no error. In 1995 these

---

[6]In fact, they confined themselves to the so-called semi-stable elliptic curves; this, however, is enough to prove FLT. Subsequently, Henri Darmon reported in the *NOTICES of the American Mathematical Society* [2], that Cristophe Breuil, Brian Conrad, Fred Diamond, and Richard Taylor have announced a full proof of what Darmon refers to as the Shimura–Taniyama–Weil conjecture.

manuscripts were published under the titles "Modular elliptic curves and Fermat's Last Theorem," by Andrew Wiles, and "Ring-theoretic properties of certain Hecke algebras," by Richard Taylor and Andrew Wiles. They took up 130 pages in the prestigious mathematics research journal the *Annals of Mathematics*.

Although the Wolfskehl prize started off at the equivalent of about $2 million, it lost considerable value in the deep recession that Germany experienced in the 1930s. However, in 1997, when presented to Wiles, it was worth a healthy $50,000.

But, from the international mathematical community's perspective, Wiles missed out on the big prize — the Fields Medal. The will of John Charles Fields, a Canadian mathematician, established a fund to provide a medal that would play the role of the Nobel Prize in mathematics. The International Congress of Mathematicians in 1932 adopted Fields' proposal, and the medal was first awarded at the next congress in 1936. It is almost certain that Wiles would have been awarded a Fields Medal. However, there is a restriction to the award. A recipient has to be under 40, and Wiles was 41 when he and Taylor completed the proof of Fermat's Last Theorem! However, the International Congress of Mathematicians, held in Berlin in 1998, awarded Wiles a special International Mathematical Union

★ silver plaque in recognition of his work.

● ● ● **BREAK 9**

(1) Sally Jones bought some pigs, goats, and sheep. Altogether she purchased 100 animals and spent $600. Now, the pigs cost $21 each, the goats $8, and the sheep $3. If there was an even number of pigs, how many of each animal did Sally buy?

(2) Show that the equation

$$x^4 + 251 = 5y^4$$

has no integral solutions for x and y.

(3) Show that the equation

$$x^2 + y^2 = 80z + 102$$

has no integral solutions for x, y, and z.

(4) How old was Diophantus when he died? (See the inscription from his tomb at the beginning of Section 3.)

(5) Show that the equation

$$x^n + y^n = z^n, \quad n \geq 3$$

has no solution in positive rational numbers x, y, z.

(6) Show that the equation $x^{n/2} + y^{n/2} = z^{n/2}$, where n is a positive integer $\neq 1, 2,$ or 4, has no solution in positive integers x, y, z.

Significant Dates for Fermat's Last Theorem (FLT)

1636?   Fermat "proves" FLT for $n = 3$.

1637?   Fermat writes a marginal note.

1640   Fermat proves FLT for $n = 4$.

1753   Euler proves FLT for $n = 3$.

1825?   Germain considers the case where p and $2p + 1$ are both prime.

1825   Dirichlet and Legendre independently prove FLT for $n = 5$.

1839   Lamé proves FLT for $n = 7$.

1847   Lamé tries to prove FLT for all n assuming unique factorization. Kummer points out Lamé's error and proves FLT for all regular primes n.

1908   Dickson proves FLT for all $n \leq 7000$.

1908   Establishment of Wolfskehl prize.

1955   Beginnings of the Shimura–Taniyama Conjecture.

1985   Frey and Serre link the Shimura–Taniyama Conjecture with FLT via a potentially non-modular elliptic curve.

1986   Ribet completes the connection between the Shimura–Taniyama Conjecture and FLT.

1988   Miyaoka claims a proof of FLT.

1993   Wiles announces a proof of FLT that turns out to be incomplete.

1994   Wiles and Taylor complete Wiles' 1993 proof of FLT by replacing the defective part of the argument.

REFERENCES

1. Cox, David, Introduction to Fermat's Last Theorem, *Amer. Math. Monthly*, **101** (1994), 3–14.
2. Darmon, Henri, A proof of the full Shimura–Taniyama–Weil conjecture is announced, *NOTICES of the American Mathematical Society*, **46** (1999), 1397–1401.
3. Gouvêa, Fernando Q., "A Marvellous Proof," *Amer. Math. Monthly*, **101** (1994), 203–222.
4. Hilton, Peter, Derek Holton, and Jean Pedersen, *Mathematical Reflections — In a Room with Many Mirrors*, Springer-Verlag, New York, 2nd printing, 1998.

5. Mazur, Barry, Number Theory as Gadfly, *Amer. Math. Monthly*, **98** (1991), 593–610.

6. Penrose, Roger, *The Emperor's New Mind*, Vintage, London, 1989.

7. Ribenboim, Paulo, *Fermat's Last Theorem for Amateurs*, Springer-Verlag, 1999.

8. Singh, Simon, *Fermat's Last Theorem*, Fourth Estate, London, 1998.

9. van der Poorten, Alf, *Notes on Fermat's Last Theorem*, Wiley, Chichester, 1996.

Information on the web can be found regarding

- Fermat's Last Theorem at `http://www.ams.org/new-in-math/fermat.html` and `http://www.ams.org/mathweb/mi-mathbytopic,html#fermat`
- Andrew Wiles at `http://www.sigmaxi.org/prizes&awards/awiles.html` and `http://www.auburn.edu/~wakefju/wiles.html`
- The Fields Medal at `http://www.math.utoronto.ca/fields`

3

CHAPTER

Fibonacci and Lucas Numbers: Their Connections and Divisibility Properties

3.1 INTRODUCTION: A NUMBER TRICK AND ITS EXPLANATION

In [1] we presented a discussion of some properties of Fibonacci and Lucas numbers, motivated by a number trick that usually intrigues an audience of students. We repeat the number trick here.

You ask a group of students to write down the numbers from 0 to 9 in a column, while doing the same thing yourself on the blackboard. You then select one student to come to the board and (with your back turned so that you can't see the board) have the student put one number against 0 and another number against 1. The student may choose to write any two numbers (but you might suggest two fairly small positive integers just to avoid tedious arithmetic). Instruct each member of the class to write the same pair of numbers next to 0 and 1 on their own paper. At this point have the student at the board erase the two numbers written on the board. You may then turn around and allow the student who has been assisting you to return to his or her place. Next instruct the class to write against 2 the sum of the entries against 0 and 1; and then write against 3 the sum of the entries against 1 and 2; and so on. When it is clear that they are all well embarked on this process, producing entries against each number from 0 to 9, you suggest that, as a check, they call out the entry against the number 6, since, as you are careful to explain, the trick would fall flat if they got different answers at the end of the process. Thus their table (which, of course, you do not see) might look like this:

0	4
1	5
2	9
3	14
4	23
5	37
6	60 (This one is used as a check)
7	97
8	157
9	254

You now ask the students to add all the entries in the second column, while you quietly write 660 on the blackboard below the 9th entry (they generally won't notice you doing this, since they will be busy calculating). When they look up they will see that you have the correct total.

They will want to know how you did your calculation so easily — indeed, how you did it at all, since you *don't know their original two numbers*. Well, let's look at the procedure from an algebraic viewpoint. If you had started with any numbers a and b, your table would have looked like Table 1. And if we supplement Table 1 with the running sums we get Table 2.

Now we see why the trick works — the running sum $\sum_9$ is actually 11 times u_6, whatever numbers a, b we choose (but the values of a and b themselves are not known when you give the answer)! However, you may have noticed, in the tables above, certain sequences of numbers occurring

0	a		
1			b
2	a	$+$	b
3	a	$+$	$2b$
4	$2a$	$+$	$3b$
5	$3a$	$+$	$5b$
6	$5a$	$+$	$8b$
7	$8a$	$+$	$13b$
8	$13a$	$+$	$21b$
9	$21a$	$+$	$34b$

TABLE 1

as coefficients, namely, 0, 1, 1, 2, 3, 5, 8, $\cdots$. These numbers are called the Fibonacci numbers (more precisely defined below). There is a companion sequence of numbers (also more precisely defined below) called the Lucas numbers, namely, 2, 1, 3, 4, 7, 11, $\cdots$. As you will see, when we study both sequences of numbers together, the mathematics is far richer than if we studied either sequence by itself.

N	u_N			$\sum_{n=0}^{N} u_n$		
0	a			a		
1			b	a	$+$	b
2	a	$+$	b	$2a$	$+$	$2b$
3	a	$+$	$2b$	$3a$	$+$	$4b$
4	$2a$	$+$	$3b$	$5a$	$+$	$7b$
5	$3a$	$+$	$5b$	$8a$	$+$	$12b$
6	$5a$	$+$	$8b$	$13a$	$+$	$20b$
7	$8a$	$+$	$13b$	$21a$	$+$	$33b$
8	$13a$	$+$	$21b$	$34a$	$+$	$54b$
9	$21a$	$+$	$34b$	$55a$	$+$	$88b$

TABLE 2

Of course, in [1] we carried the explanation of the number trick much further, so that it occupied the second and third sections of Chapter 3 of [1] — and, in fact, that discussion eventually led us to some properties involving divisibility.

Here we want to proceed directly to those and other *divisibility* properties of these numbers (see Section 3.3 of [1]). However, we will devote a final section of this chapter to a nice, simplifying observation about polynomial identities relating Fibonacci and Lucas numbers.

Let us first recall some basic facts which may be found in Chapter 3 of [1].

The Fibonacci and Lucas sequences of integers, written $\{F_n\}$ and $\{L_n\}$, respectively, both satisfy the recurrence relation

$$u_{n+2} = u_{n+1} + u_n, \quad n \geq 0 \tag{1}$$

For the Fibonacci sequence, the initial conditions are

$$F_0 = 0, \quad F_1 = 1 \tag{2}$$

while, for the Lucas sequence, the initial conditions are

$$L_0 = 2, \quad L_1 = 1 \tag{3}$$

(Why do you think we don't consider the sequence $\{u_n\}$ with initial conditions $u_0 = 1, u_1 = 1$?)

Here is a table of the values of F_n and L_n up to $n = 13$. You may find it useful to extend it further in order to verify some of the following Theorems in special cases.

n	0	1	2	3	4	5	6	7	8	9	10	11	12	13
F_n	0	1	1	2	3	5	8	13	21	34	55	89	144	233
L_n	2	1	3	4	7	11	18	29	47	76	123	199	322	521

TABLE 3

Now let α, β be the roots of the equation $x^2 - x - 1 = 0$, so that

$$\alpha + \beta = 1, \quad \alpha\beta = -1 \tag{4}$$

Then the Fibonacci and Lucas numbers are given by the **Binet formulae**

$$F_n = \frac{\alpha^n - \beta^n}{\alpha - \beta}, \quad L_n = \alpha^n + \beta^n \tag{5}$$

From these formulae we infer our primary divisibility results,[1] namely,

Theorem 1 *If $m|n$, then $F_m|F_n$.*

Theorem 2 *If $m|n$ oddly, then $L_m|L_n$.*

Here, and subsequently, ***m|n oddly*** if m divides n with *odd* quotient.

These theorems are refined in the following results. Let $m, n \geq 1$ and $\gcd(m, n) = d$. Then

Theorem 3 $\gcd(F_m, F_n) = F_d$.

To state the next theorem, we need a definition. If m is a positive integer, then $|m|_2$ is the highest power of 2 dividing m; we may call it the **2-*value*** of m. Now comes the theorem.

Theorem 4

$$\gcd(L_m, L_n) = \begin{cases} L_d & if \ |m|_2 = |n|_2 \\ 1 & if \ |m|_2 \neq |n|_2 \quad and \quad 3 \nmid d \\ 2 & if \ |m|_2 \neq |n|_2 \quad and \quad 3|d \end{cases}$$

Theorems 1, 2, 3, 4 may all be found in [1]. But there is an important companion result to Theorems 3 and 4 not mentioned in [1]; we will be using it here. The proofs of Theorems 3, 4, 5 may be found in [2].

Theorem 5

$$\gcd(F_m, L_n) = \begin{cases} L_d & if \ |m|_2 > |n|_2 \\ 1 & if \ |m|_2 \leq |n|_2 \quad and \quad 3 \nmid d \\ 2 & if \ |m|_2 \leq |n|_2 \quad and \quad 3|d \end{cases}$$

An easy result which helps to explain the last parts of Theorems 4 and 5 states that

$$F_n \text{ even} \quad \Leftrightarrow \quad L_n \text{ even} \quad \Leftrightarrow \quad 3|n \tag{6}$$

---

[1]The symbol $m|n$ means m divides n, that is, m is a factor of n. The symbol $m \nmid n$ means m does not divide n.

• • • **BREAK 1**

(1) Show, using the Binet formulae and (4), that

$$L_{n+1}L_n = L_{2n+1} + (-1)^n$$

and

$$F_{n+1}L_n = F_{2n+1} + (-1)^n$$

(Notice the enrichment of the mathematics, of which we spoke earlier.)

(2) Try to prove (6). (Hint: Look at the first few terms of the Fibonacci and Lucas sequences mod 2.)

As pointed out in [1], given any positive integer q, there is a Fibonacci number F_m, with $m \geq 1$, such that $q|F_m$. If m is taken as small as possible, we call it the **Fibonacci Index** of q and write

$$m = FI(q) \tag{7}$$

On the other hand, there are positive integers q that exactly divide no Lucas number L_n; 5 and 8 are examples. If there exists $m \geq 1$ such that $q|L_m$, we call q **Lucasian**; and, if m is taken as small as possible, we call it the **Lucas Index** of q and write

$$m = LI(q) \tag{8}$$

We are now ready to start proving a remarkable series of results related to these ideas. Note that, from now onwards, we will only be considering F_n and L_n for[2] $n \geq 1$.

3.2   A FIRST SET OF RESULTS ON THE FIBONACCI AND LUCAS INDICES

We start with some fairly easy (but important) results.

Theorem 6   *Let $q \geq 1$ and $FI(q) = m$. Then, for any $n \geq 1$,*

$$q|F_n \quad \Leftrightarrow \quad m|n \quad \Leftrightarrow \quad F_m|F_n$$

In particular, if $q_1|q_2$, then $FI(q_1)|FI(q_2)$.

---

[2]The proofs of Theorems 3, 4, 5 given in [2] involve the use of Fibonacci and Lucas numbers F_n, L_n with n negative.

Proof   That $m|n$ implies $F_m|F_n$ is Theorem 1. That $F_m|F_n$ implies $q|F_n$ is obvious, since $q|F_m$. Thus we are left to prove that $q|F_n$ implies that $m|n$. Now, if $q|F_n$, then, since we also have $q|F_m$, it follows that $q|\gcd(F_m, F_n)$ or, by Theorem 3, $q|F_d$, where $d = \gcd(m, n)$. But m is the *smallest* positive integer such that $q|F_m$, so $d = m$, or, equivalently, $m|n$. $\square$

We leave the proof of the last statement of the theorem as an exercise, and turn to the corresponding statement for the Lucas Index. We assume all the numbers involved to be *Lucasian*, so that they do have a Lucas Index, but we further assume them to be ≥ 3. We have then

Theorem 7   *Let $q \geq 3$ be Lucasian with $LI(q) = m$. Then, for any $n \geq 1$,*

$$q|L_n \quad \Leftrightarrow \quad m|n \text{ oddly} \quad \Leftrightarrow \quad L_m|L_n$$

In particular,

if $q_1, q_2 \geq 3$ are Lucasian, and if $q_1|q_2$, then $LI(q_1)|LI(q_2)$ oddly.

Proof   As for Theorem 6, the key step in the proof is to establish that

$$q|L_n \quad \Rightarrow \quad m|n \text{ oddly.}$$

We have $q|L_m$, $q|L_n$, so $q|\gcd(L_m, L_n)$. It is here that we need the hypothesis $q \geq 3$; for Theorem 4 then tells us that we must have $|m|_2 = |n|_2$ and $\gcd(L_m, L_n) = L_d$. It now follows as in Theorem 6 that $d = m$, $m|n$; but, since $|m|_2 = |n|_2$, we conclude that $m|n$ oddly.

We again leave the proof of the last statement of the theorem as an exercise. $\square$

● ● ● **BREAK 2**

(1) Prove the last part of Theorem 6.
(2) Similarly, prove the last part of Theorem 7.
(3) Use Theorem 6 to infer that any Fibonacci number divisible by 4 is divisible by 8. Find a similar statement for Lucas numbers that follows from Theorem 7.
(4) Prove that 5 and 8 are *not* Lucasian. (Hint: Record the residue classes mod 5 and mod 8 of the sequence of Lucas numbers.)

We now use Theorem 5 to prove a sensational — and very recent — result (first published in [5] in 1997).[3]

Theorem 8 *No Fibonacci number > 3 is Lucasian.*

Remark We have already noted that 5, 8 are not Lucasian.

Proof of Theorem 8 We prove Theorem 8 in the form: if $F_m \geq 3$ is Lucasian, then $F_m = 3$. We first prove that if $F_m \geq 3$ is actually a Lucas number, then $F_m = 3$.

So we suppose $F_m = L_n \geq 3$. Then $\gcd(F_m, L_n) = L_n \geq 3$ so, by Theorem 5, $|m|_2 > |n|_2$ and $\gcd(F_m, L_n) = L_d$, where

$$d = \gcd(m, n)$$

But, for $n \geq 1$, L_n is an increasing function of n (why?), so that

$$L_d = L_n \quad \Leftrightarrow \quad d = n \quad \Leftrightarrow \quad n|m$$

But, since $|m|_2 > |n|_2$, $n|m$ implies that $2n|m$. Hence we have

$$L_n | F_{2n} | F_m$$

since $F_{2n} = F_n L_n$ (from the Binet formulae (5)). But $L_n = F_m$, so $L_n = F_{2n}$, whence $F_n = 1$. This implies that $n = 1$ or 2. However, $n = 1$ is impossible, since $L_n \geq 3$ and $L_1 = 1$. Thus $n = 2$ and $F_m = L_2 = 3$ (so $m = 4$). Thus we have proved that, if $F_m \geq 3$ is a Lucas number, $F_m = 3$.

We now suppose that $F_m \geq 3$ is Lucasian, say $F_m|L_n$. Then $\gcd(F_m, L_n) = F_m \geq 3$. A second application of Theorem 5 shows us that $F_m = L_d \geq 3$, where $d = \gcd(m, n)$. But we have already seen that this implies that $F_m = 3$. Our proof is complete. □

Of course, there are non-Lucasian numbers that are not Fibonacci numbers, for any multiple of a non-Lucasian number is non-Lucasian. We will find in Section 4 another source of non-Lucasian numbers.

3.3 ON ODD LUCASIAN NUMBERS

We will prove in this section the remarkable fact that if s is an *odd* Lucasian number, then all powers of s are Lucasian. This is intrinsically surprising;

[3]It is striking that new results are to be found today about number sequences that have been studied for 600 years and 100 years, respectively.

but it becomes even more surprising when one realizes that it is false *for every even number.* For $2, 4$ are Lucasian but 8 is not; thus the cube of every even number is non-Lucasian. Here is a truly unexpected situation.

We establish the facts for odd Lucasian numbers by means of a key formula which we are about to prove, namely, for any odd integer s, and any positive integer r,

$$L_{sr} = \pm s L_r + c L_r^3, \quad \text{for some integer } c.$$

To obtain this formula, we first prove a fairly deep result for Lucas numbers L_n with n even.

Theorem 9

$$L_{2kr} \equiv \begin{cases} 2 \bmod L_r^2 & \text{if } r \text{ is odd} \\ 2(-1)^k \bmod L_r^2 & \text{if } r \text{ is even} \end{cases}$$

Proof By the Binet formula for L_n, together with $\alpha\beta = -1$, we see that

$$L_{2r} = \alpha^{2r} + \beta^{2r} = (\alpha^r + \beta^r)^2 - 2\alpha^r \beta^r = L_r^2 - 2(-1)^r,$$

so

$$L_{2r} \equiv \begin{cases} 2 \bmod L_r^2 & \text{if } r \text{ is odd} \\ -2 \bmod L_r^2 & \text{if } r \text{ even} \end{cases} \tag{9}$$

Hence $L_{4r} = L_{2r}^2 - 2$, so, by (9),

$$L_{4r} \equiv 2 \bmod L_r^2 \tag{10}$$

Thus the statement of Theorem 9 holds good if $k = 1$ or 2. We may therefore prove Theorem 9 by using a rather unusual kind of inductive argument; that is, by assuming the truth of the statement of Theorem 9 with k replaced by $k-2$ and by $k-1$ ($k \geq 3$) and proving its truth in the form given. The formula we use to do this, namely,

$$L_{2kr} + L_{2(k-2)r} = L_{2(k-1)r} L_{2r} \tag{11}$$

is based, as before, on the Binet formula for L_n and the identity $\alpha\beta = -1$. We suggest that you verify (11) as an exercise, starting from the right-hand side and applying the Binet formula for L_n; in fact, (11) is a special case of the fundamental quadratic identity (24), which you will meet in Section 6.

Let us first assume r odd. Then we know, by the inductive hypothesis, that $L_{2(k-2)r} \equiv 2 \bmod L_r^2$, $L_{2(k-1)r} \equiv 2 \bmod L_r^2$, and, by (9), that $L_{2r} \equiv 2 \bmod L_r^2$. Thus (11) tells us that

$$L_{2kr} \equiv 4 - 2 = 2 \bmod L_r^2$$

as claimed. Next assume that r is even. Then

$$L_{2(k-2)r} \equiv 2(-1)^k \bmod L_r^2$$
$$L_{2(k-1)r} \equiv 2(-1)^{k-1} \bmod L_r^2$$
$$L_{2r} \equiv -2 \bmod L_r^2$$

Thus, by (11),

$$L_{2kr} \equiv -4(-1)^{k-1} - 2(-1)^k = 2(-1)^k \bmod L_r^2$$

as claimed, and the proof of Theorem 9 is complete. $\qquad\square$

● ● ●   **BREAK 3**

(1) Verify Theorem 9 (i) for $r = 3, k = 3$; (ii) for $r = 4, k = 3$.
(2) Verify formula (11).

Now we introduce an odd number $s = 2k + 1$. We know from Theorem 2 that $L_r | L_{sr}$. We will prove

Theorem 10   *Let $s = 2k + 1$. Then*

$$\frac{L_{sr}}{L_r} \equiv \begin{cases} s \bmod L_r^2 & \text{if } r \text{ is odd} \\ (-1)^k s \bmod L_r^2 & \text{if } r \text{ is even} \end{cases}$$

Proof   Let us first assume r *odd*. By elementary algebra we know that

$$\begin{aligned}
\frac{L_{sr}}{L_r} &= \frac{\alpha^{sr} + \beta^{sr}}{\alpha^r + \beta^r} \\
&= \alpha^{2kr} - \alpha^{(2k-1)r}\beta^r + \alpha^{(2k-2)r}\beta^{2r} + \cdots + (-1)^k \alpha^{kr}\beta^{kr} \\
&\quad + \cdots + \alpha^{2r}\beta^{(2k-2)r} - \alpha^r \beta^{(2k-1)r} + \beta^{2kr}
\end{aligned} \tag{12}$$

Collecting terms from either end of (12), we obtain (remember that $\alpha\beta = -1$ and r is odd)

$$\frac{L_{sr}}{L_r} = L_{2kr} + L_{2(k-1)r} + L_{2(k-2)r} + \cdots + L_{2r} + 1$$

$$\equiv (2k + 1) \bmod L_r^2, \quad \text{by Theorem 9,}$$

$$= s \bmod L_r^2$$

Now assume r *even*. We again proceed to (12) and collect terms as before. Now, however, with r even, we obtain

$$\frac{L_{sr}}{L_r} = L_{2kr} - L_{2(k-1)r} + L_{2(k-2)r} - \cdots - (-1)^{k-1}L_{2r} + (-1)^k$$

$$\equiv (-1)^k(2k + 1) \bmod L_r^2, \quad \text{by Theorem 9,}$$

$$= (-1)^k s \bmod L_r^2$$

$\square$

Except to verify the formula of Theorem 10, we will not be concerned in the sequel with the precise sign[4] in Theorem 10. Thus we will be content to quote: if s is odd, then

$$\frac{L_{sr}}{L_r} \equiv \pm s \bmod L_r^2 \tag{13}$$

Indeed, we go further. We know from (13) that

$$\frac{L_{sr}}{L_r} = \pm s + cL_r^2$$

for some integer c, so that, as claimed earlier,

$$L_{sr} = \pm sL_r + cL_r^3 \tag{14}$$

It is (14) that we regard as our *key formula*. It enables us to prove yet another remarkable theorem. To state this theorem we introduce some notation. First,

$$s^k[]\ell$$

means that $s^k | \ell$ but $s^{k+1} \nmid \ell$; if $s^k[]\ell$ and s is a prime p, we may also write[5]

$$|\ell|_p = k$$

---

[4]But notice that we could not have proved Theorem 10, even with $\pm$ replacing the precise sign, without knowing the precise sign in Theorem 9. Don't get sloppy too soon — that would be unlicensed sloppiness!

[5]The notation $|\ell|_p$ is consistent with the meaning earlier attached to the symbol $|m|_2$.

Second,

$$s^k||\ell$$

means that $s^k|\ell$ with quotient prime to s. Obviously, $s^k[]\ell$ if $s^k||\ell$, but the converse only holds if s is a prime. We now prove a result that has not been published before.

Theorem 11   *Suppose that s is odd and $s|L_r$. Then, for any $q \geq 1$,*

(i) $s^q|L_r \Leftrightarrow s^{q+1}|L_{sr}$,
(ii) $s^q[]L_r \Leftrightarrow s^{q+1}[]L_{sr}$,
(iii) $s^q||L_r \Leftrightarrow s^{q+1}||L_{sr}$.

Before proving this (rather surprising) theorem, we give some examples to illustrate it, which you might like to check; and we make a remark harking back to the claim we made at the start of this section.

Examples of Theorem 11   Using $L_{18} = 5778 = 27 \cdot 214$ you may
    verify that

$$3|L_2, \quad 3^2|L_6, \quad 3^3|L_{18}$$
$$3[]L_2, \quad 3^2[]L_6, \quad 3^3[]L_{18}$$
$$3||L_2, \quad 3^2||L_6, \quad 3^3||L_{18}$$

Next we remark that it is part (i) of this theorem that establishes the fact that every power of s is Lucasian if s is Lucasian. For it follows immediately from (i) that

$$s|L_r \Rightarrow s^q|L_{s^{q-1}r}, \quad q \geq 1 \tag{15}$$

We now come to the proof[6] of Theorem 11.

Proof of Theorem 11   The entire proof is based on (14). We begin by
    proving the $\Rightarrow$ part of (i). For if $s^q|L_r$, then $s^{q+1}|sL_r$ and $s^{3q}|cL_r^3$.
    Thus, since $q \geq 1$, $s^{q+1}|cL_r^3$, so, by (14), $s^{q+1}|L_{sr}$. Next, we prove
    the $\Rightarrow$ part of (ii). If $s^q[]L_r$, then $s^{q+1}[]sL_r$, but $s^{q+2}|cL_r^3$. It follows,
    from (14), that $s^{q+1}[]L_{sr}$. Next we prove the $\Rightarrow$ part of (iii). If $s^q||L_r$,
    then $s^{q+1}||sL_r$. Since $s^{q+2}|cL_r^3$, it follows, from (14), that the quotient

---

[6]It is obvious that, in practice, the $\Rightarrow$ parts of Theorem 11 would be used far more often than the $\Leftarrow$ parts.

of L_{sr} by s^{q+1} is the sum of a number prime to s and a number divisible by s, and hence is prime to s. Thus $s^{q+1} || L_{sr}$.

We now prove the reverse implications, starting with $\Leftarrow$ in (ii). Since $s | L_r$, there exists $t \geq 1$ such that $s^t [] L_r$. But then $s^{t+1} [] L_{sr}$, so $t + 1 = q + 1, t = q$. We next prove $\Leftarrow$ in (i). Given $s^{q+1} | L_{sr}$, suppose that $s^{t+1} [] L_{sr}, t \geq q \geq 1$. Then $s^t [] L_r$, so $s^q | L_r$. Finally, we prove $\Leftarrow$ in (iii). Given $s^{q+1} || L_{sr}$, we know that $s^q [] L_r$ (why?). Thus $s^{q+1} || L_{sr}, s^{q+2} | cL_r^3$, so, by (14), $s^{q+1} || s L_r$, and finally, $s^q || L_r$. $\quad\square$

We now make another crucial use of our key formula (14).

Theorem 12   *Let p be a prime divisor of L_r. Then, with the data of Theorem 11, that is, assuming that s is odd and $s | L_r$, we conclude that*

$$|L_{sr}|_p = |L_r|_p + |s|_p \qquad (16)$$

Proof   Let $|L_r|_p = n \geq 1$, so $|cL_r^3|_p \geq 3n$; and, repeating (14),

$$L_{sr} = \pm s L_r + c L_r^3$$

Now, $|s|_p \leq n$, since $s | L_r$, so $|s L_r|_p \leq 2n < |cL_r^3|_p$, yielding

$$|L_{sr}|_p = |\pm s L_r + c L_r^3|_p = |s L_r|_p = |L_r|_p + |s|_p$$

$$\square$$

We use Theorem 12 to prove one of our most important results on the Lucas Index.

Theorem 13   *Suppose s is odd and $s^q || L_r$, so that $s^{q+1} || L_{sr}$, where $q \geq 1$. Then, if $r = LI(s^q)$,*

$$sr = LI(s^{q+1})$$

Proof   The claim is trivial if $s = 1$, so we may assume $s \geq 3$ (remember that s is odd). By Theorem 7, with $LI(s^{q+1}) = t$, we have $r | t | sr$ (with odd quotients). It follows that $t = ur$, with $u | s$. If $u \neq s$, then there exists a prime p such that $p | s$ and $|u|_p < |s|_p$. Now, by Theorem 12, since $p | L_r$, we have

$$|L_{ur}|_p = |L_r|_p + |u|_p < |L_r|_p + |s|_p = |L_{sr}|_p = (q+1)|s|_p,$$

since $s^{q+1} || L_{sr}$. Thus $s^{q+1} \nmid L_{ur}$, contradicting $LI(s^{q+1}) = ur$. Hence $u = s$ and $LI(s^{q+1}) = sr$, as claimed. $\quad\square$

Remarks on Theorem 13

(i)   The hypothesis $s^q||L_r$ is essential; it is not sufficient to suppose that $s^q[\,]L_r$; indeed, if $s^q[\,]L_r$, then the conclusion may be false. See [3] for detailed results on this issue.

(ii) Of course, we may iterate Theorem 13 to infer that

$$s^n r = LI(s^{q+n})$$

(iii) The converse of Theorem 13 holds. If $s^q||L_r$ and $sr = LI(s^{q+1})$, let $LI(s^q) = t$. Then $t|r$ oddly, so $L_t|L_r$, whence $s^q||L_t$. Thus, by Theorem 13, $LI(s^{q+1}) = st$, so that $st = sr, t = r$.
However, it is unlikely that the converse would be useful in determining the Lucas Index.

Example 14   We observe that $L_{15} = 1364 = 4 \times 11 \times 31$. Now $L_3 = 4$, $L_5 = 11$. Thus

$$LI(31) = 15 \quad \text{and} \quad 31||L_{15}$$

It follows that $LI(31^{n+1}) = 15 \cdot 31^n$, and that any Lucas number divisible by 31 is divisible by 1364. (Why?)

● ● ●   **BREAK 4**

(1) Answer the last question above.
(2) Find another example similar to that given.

3.4   A THEOREM ON LEAST COMMON MULTIPLES

We suppose $a, b \geq 3$ are Lucasian and ask: is $\text{lcm}(a, b)$ necessarily Lucasian? The answer is no, for 3 and 7 are Lucasian, but 21 is not. The complete answer is to be found in the following theorem.

Theorem 15   *Let $a, b \geq 3$ and let a, b be Lucasian with $LI(a) = q$, $LI(b) = r$. Then $\text{lcm}(a, b)$ is Lucasian $\Leftrightarrow |q|_2 = |r|_2$.*
Moreover, if $|q|_2 = |r|_2$, then

$$LI(\text{lcm}(a, b)) = \text{lcm}(LI(a), LI(b)) \tag{17}$$

Proof   Set $c = \text{lcm}(a, b)$ and suppose c Lucasian. Then $a|c$ so, by Theorem 7, $q|LI(c)$ oddly. Similarly, $r|LI(c)$ oddly. Hence

$$|q|_2 = |LI(c)|_2 = |r|_2$$

Conversely, suppose $|q|_2 = |r|_2$ and let $\ell = \text{lcm}(q, r)$. Then $q|\ell$ oddly, so $a|L_\ell$. Similarly, $b|L_\ell$, so $c|L_\ell$ and c is Lucasian.

Finally, assume $|q|_2 = |r|_2$ and $LI(c) = m$. Then $m|\ell$ oddly. On the other hand, $a|c|L_m$, so $q|m$ oddly, and similarly $r|m$ oddly, so $\ell|m$ oddly, whence $\ell = m$. But this is exactly (17).   □

Example 16 $LI(3) = 2$, $LI(7) = 4$, so Theorem 15 gives a second explanation of the fact that the Fibonacci number 21 is not Lucasian. On the other hand, $LI(9) = 6$, $(L_6 = 18)$, $LI(41) = 10$ $(L_{10} = 123)$, and $|6|_2 = |10|_2 = 1$. Thus 369 is Lucasian and $LI(369) = 30$.

● ● ●   BREAK 5

Of course, the analogue of (17) holds for the Fibonacci Index with no restrictions on a and b. Prove this; that is, prove that

$$FI(\text{lcm}(a, b)) = \text{lcm}(FI(a), FI(b)) \qquad (18)$$

3.5   THE RELATION BETWEEN THE FIBONACCI AND LUCAS INDICES

In this section we establish a precise relation between the Fibonacci and Lucas indices of a Lucasian number $q \geq 3$. The theorem states:

Theorem 17   *If q is Lucasian, and $q \geq 3$, then*

$$FI(q) = 2LI(q) \qquad (19)$$

Notice that (19) does *not* hold if $q = 1$ or 2. For $FI(1) = LI(1) = 1$, $FI(2) = LI(2) = 3$. In proving (19) we will need a nice fact from elementary number theory. We leave the proof of the following lemma to the reader.

Lemma 18   *Let $q|ab$. Then q can be written as $q = uv$, where $u|a$, $v|b$.*

Proof of Theorem 17   We first assume that $q \geq 3$ is odd. Suppose $FI(q) = m$, $LI(q) = n$. Then $q|L_n$, so $q|F_{2n}$ and $m|2n$. On the other hand, $q \nmid F_n$, since $\gcd(F_n, L_n) = 1$ or 2 (see Theorem 5). Thus $m \nmid n$, whence m is even, say $m = 2c$, $c|n$ (oddly). Now $q|F_c L_c$, since $q|F_{2c}$, so, by our lemma, $q = uv$, where $u|F_c$, $v|L_c$. Then $u|F_c|F_n$ and $u|q|L_n$, so, by Theorem 5, u, being odd, must be 1. Thus $q = v$, so $q|L_c$, whence $n|c$ (oddly), by Theorem 7. This shows that $n = c$, so $m = 2n$, as claimed.

We next prove (19) for $q = 2s$, where s is odd, $s \geq 3$. For let $FI(s) = m, LI(s) = n$, so that $m = 2n$. Now, if $3|n$, then $2|L_n$, so $2s|L_n$, and similarly $2s|F_m$, so $m = FI(2s), n = LI(2s)$, and (19) holds. If $3 \nmid n$, then $2|L_{3n}$ and $s|L_{3n}$, so $2s|L_{3n}$. Moreover, $3n = LI(2s)$, since $LI(2s)$ must be a number t such that $n|t$ oddly, $3|t$ and $t|3n$ oddly. Even more obviously, $FI(2s) = 3m$ if $3 \nmid m$ (equivalent, of course, to $3 \nmid n$), so (19) holds in this case.

We next prove (19) for $q = 4s$, where s is odd, $s \geq 3$. We first observe that $4s$ is Lucasian $\Leftrightarrow LI(s)$ is odd, by Theorem 15, since $LI(4)$ is odd. Again, by Theorem 15, if $4s$ is Lucasian,

$$LI(4s) = \text{lcm}(3, LI(s))$$

Thus

$$
\begin{aligned}
FI(4s) &= \text{lcm}(FI(4), FI(s)) \\
&= \text{lcm}(6, 2LI(s)) \\
&= 2\,\text{lcm}(3, LI(s)) = 2LI(4s);
\end{aligned}
$$

and, finally, $FI(4) = 2LI(4) = 6$.

We claim that the proof of Theorem 17 is now complete. For 8 is not Lucasian, so no number divisible by 8 is Lucasian. Thus our analysis of odd numbers s, of even numbers $2s$ (s odd), and even numbers $4s$ (s odd), has included all Lucasian numbers $q \geq 3$. We rest our case! $\square$

• • • BREAK 6

(1) Show that, if $4s$ is Lucasian, then $LI(4s) = LI(2s)$.
(2) Prove Lemma 18. (If you can't prove it, illustrate it with examples.)

3.6 ON POLYNOMIAL IDENTITIES
RELATING FIBONACCI AND LUCAS NUMBERS

There is an extensive literature in which polynomial identities relating the Fibonacci and Lucas numbers are discovered and proved[7] (see, for example, [4, 6]). Of course, there is no problem in proving a *linear* identity;

---

[7]In this section we regard F_n and L_n as defined for all integer values of n.

for a linear identity may be established by simply verifying it for two successive values of n. For example, if we want to verify that

$$L_n = F_{n+1} + F_{n-1}, \quad \text{for all } n, \tag{20}$$

we simply observe that this holds for $n = 1$ ($1 = 1 + 0$), and for $n = 2$ ($3 = 2 + 1$). This works because the defining recurrence relation for both Fibonacci and Lucas numbers is the *same linear* relation:

$$u_{n+2} = u_{n+1} + u_n \tag{21}$$

The companion relation to (20) is

$$5F_n = L_{n+1} + L_n \tag{22}$$

and the remaining key linear relations[8] are

$$F_{-n} = (-1)^{n-1}F_n, \quad L_{-n} = (-1)^n L_n \tag{23}$$

But we may regard any linear relation as trivial if true, by the argument above, and proceed to discuss relations of higher degree. First, however, we take a break.

● ● ● **BREAK 7**

(1) Prove (22) for all values of n.
(2) Use the recurrence relation to carry Table 1 back one place; and thus prove the relations in (23).
(3) Explain why linear identities involving linear second order recurrence relations (and not only (21)) may be proved in this easy fashion.

Let us now turn to quadratic identities. We first prove the identity

$$L_m L_n = L_{m+n} + (-1)^n L_{m-n} \tag{24}$$

(Notice that (11) is a special case of (24).)

To prove (24) we use the Binet identity (5). Thus (with m, n *any* integers)

$$\begin{aligned}
L_m L_n &= (\alpha^m + \beta^m)(\alpha^n + \beta^n) \\
&= \alpha^{m+n} + \beta^{m+n} + \alpha^n \beta^n (\alpha^{m-n} + \beta^{m-n}) \\
&= L_{m+n} + (-1)^n L_{m-n}, \quad \text{since} \quad \alpha\beta = -1
\end{aligned}$$

---

[8]There is a small extra subtlety about proving the identities (23), due to the presence of a sign depending on n, but it does not invalidate the proof strategy we have described.

There are two other quadratic identities similar in nature to (24); they are

$$5F_m F_n = L_{m+n} - (-1)^n L_{m-n} \tag{25}$$
$$F_m L_n = F_{m+n} + (-1)^n F_{m-n} \tag{26}$$

However, it is not difficult to see that any one of (24), (25), (26) may be derived from any other by applying (20) or (22). Thus, for example, given (22), (24), we have

$$
\begin{aligned}
5F_m L_n &= (L_{m+1} + L_{m-1})L_n \\
&= L_{m+n+1} + (-1)^n L_{m-n+1} + L_{m+n-1} + (-1)^n L_{m-n-1} \\
&= 5F_{m+n} + (-1)^n 5F_{m-n}
\end{aligned}
$$

establishing (26). Likewise, (22) and (26) together establish (25).

Now, given any conjectured polynomial identity of degree $k \geq 2$, we may use the quadratic identities (24), (25), (26) to reduce it to an equivalent identity of degree $k - 1$, and hence eventually to a linear identity. This linear identity may then be verified — or disproved — by checking it for two successive values of n.

Thus one can say that, in addition to linear identities, one only needs *one* quadratic identity to establish *all* polynomial identities relating Fibonacci and Lucas numbers — one may choose that quadratic identity to be (24), (25), or (26). Moreover, one can make the derivation process entirely algorithmic if one wishes. Let us give an example of the process.

Example 19   Prove that

$$L_i L_j L_k = L_{i+j+k} + (-1)^i L_{-i+j+k} + (-1)^j L_{i-j+k} + (-1)^k L_{i+j-k}$$

To reduce this cubic identity to a quadratic identity we invoke (24) to write it as

$$
\begin{aligned}
(L_{i+j} &+ (-1)^j L_{i-j})L_k \\
&= L_{i+j+k} + (-1)^i L_{-i+j+k} + (-1)^j L_{i-j+k} + (-1)^k L_{i+j-k}
\end{aligned}
$$

We invoke (24) again to reduce this to a linear identity. We find that we have only to verify that

$$(-1)^{j+k} L_{i-j-k} = (-1)^i L_{-i+j+k}$$

but this is an immediate consequence of the identity $L_{-n} = (-1)^n L_n$ (see (23)).

We emphasize that we have been discussing polynomial identities relating, specifically, the Fibonacci and Lucas numbers. However, the simple procedure we described for verifying a linear identity applies to any such identity involving a sequence, or combination of sequences, satisfying (21). Let us now illustrate this point, by proving the following theorem.

Theorem 20   *Let $\{u_n\}$ be an arbitrary sequence satisfying* (21). *Then*

$$u_{m+n-1} = u_m F_n + u_{m-1} F_{n-1}. \tag{27}$$

Proof   The relationship (27) has the appearance of a *quadratic* identity. However, we can convert it into a *linear* identity by holding m and the sequence $\{u_n\}$ fixed and regarding (27) as a relationship between the sequence $\{u_n\}$ and the Fibonacci sequence $\{F_n\}$, true for all n. Now, to prove (27), we have only to verify it for two successive values of n; we choose $n = 1, 2$. We recall that $F_0 = 0$, $F_1 = 1$, $F_2 = 1$.

With $n = 1$, (27) asserts that $u_m = u_m$, evidently true; with $n = 2$, (27) asserts that $u_{m+1} = u_m + u_{m-1}$, also evidently true. Thus (27) is established for all values of m, n and all sequences $\{u_n\}$ satisfying (21). $\square$

We may regard (27) as showing why, from a *mathematical* point of view, the Fibonacci sequence has a primacy[9] among all sequences satisfying (21). For it shows how to express the terms of *any* other sequence $\{u_n\}$ satisfying (21) as a linear combination of the terms of the sequence $\{F_n\}$, namely,

$$u_n = u_1 F_n + u_0 F_{n-1} \tag{28}$$

Notice the example of this in the first display of Table 2.

We will use Theorem 20 to provide alternative proofs of two of our basic divisibility results, namely, Theorem 1 and Theorem 3. In fact, we will invoke (27) with $u_n = F_n$; that is, we will exploit the relationship

$$F_{m+n-1} = F_m F_n + F_{m-1} F_{n-1} \tag{29}$$

We suppose a, b to be positive integers with $a \geq b$; then

$$a = qb + r, \quad 0 \leq r \leq b - 1 \tag{30}$$

Setting $m = a - b + 1$, $n = b$ in (29), we obtain

$$F_a = F_{a-b+1} F_b + F_{a-b} F_{b-1}$$

---

[9]Of course, the Fibonacci sequence also has a primacy in biology.

Thus

$$\gcd(F_a, F_b) = \gcd(F_{a-b}F_{b-1}, F_b) \quad \text{(why?)}$$

Now it is easy to see, from the recurrence relation $F_{n+2} = F_{n+1} + F_n$, that

$$\gcd(F_{n+2}, F_{n+1}) = \gcd(F_{n+1}, F_n).$$

Proceeding in this way we see that

$$\gcd(F_b, F_{b-1}) = \gcd(F_2, F_1) = \gcd(1, 1) = 1$$

Thus we see that F_b, F_{b-1} are coprime. It follows that

$$\gcd(F_{a-b}F_{b-1}, F_b) = \gcd(F_{a-b}, F_b) \quad \text{(why?)}$$

so that

$$\gcd(F_a, F_b) = \gcd(F_{a-b}, F_b) \tag{31}$$

Iterating (31) and using (30), we arrive at the conclusion

$$\gcd(F_a, F_b) = \gcd(F_r, F_b) \tag{32}$$

where r is the remainder on dividing a by b.

We complete the proof of Theorem 1 as follows. If $b|a$, then $r = 0$. Thus, in this case,

$$\gcd(F_a, F_b) = \gcd(F_0, F_b) = \gcd(0, F_b) = F_b$$

that is, $F_b|F_a$.

We complete the proof of Theorem 3 as follows. We employ the Euclidean algorithm to calculate $d = \gcd(a, b)$. That is, we obtain a sequence of remainders $r_1(= r), r_2, \cdots, r_k(= d), r_{k+1}(= 0)$, where r_i is the remainder when r_{i-2} is divided by r_{i-1}, $i = 1, 2, \cdots, k+1$, with $r_{-1} = a, r_0 = b$. But (32) tells us that

$$\gcd(F_a, F_b) = \gcd(F_b, F_{r_1}) = \gcd(F_{r_1}, F_{r_2})$$
$$= \cdots = \gcd(F_{r_k}, F_{r_{k+1}}) = \gcd(F_d, F_0) = F_d,$$

establishing Theorem 3. Notice that Theorems 1 and 3 have here been established without recourse to the Binet formulae.

● ● ●   **BREAK 8**

    (1) Answer the two questions (introduced by "why?") in our proof of formula (31).

    (2) Prove (29) as a consequence of our basic quadratic identity (25).

    (3) Find $\gcd(377, 10946)$.

REFERENCES

1. Hilton, Peter, Derek Holton, and Jean Pedersen, *Mathematical Reflections — In a Room with Many Mirrors*, 2nd printing, Springer Verlag NY, 1998.

2. Hilton, Peter, and Jean Pedersen, Fibonacci and Lucas numbers in teaching and research, *Journées Mathématiques & Informatique*, **3** (1991–1992), 36–57.

3. Hilton, Peter, and Jean Pedersen, On a generalization of Fibonaccian and Lucasian numbers, *Proceedings of the A. C. Aitken Centenary Conference, Otago Conference Series* **5**, University of Otago Press, Dunedin, New Zealand (1995), 153–163.

4. Hilton, Peter, and Jean Pedersen, A fresh look at old favorites: the Fibonacci and Lucas sequences revisited, *Australian Mathematical Society Gazette* **25**, No. 3 (1998), 1–15.

5. Hilton, Peter, Jean Pedersen, and Larry Somer, On Lucasian numbers, *The Fibonacci Quarterly* **35**, No. 1 (1997), 43–47.

6. Selections from Elementary (or Advanced) Problems, appearing in *The Fibonacci Quarterly*:
    Elementary Problem B-606, Vol. 25, No. 4, 1987, p. 370;
    Elementary Problem B-613, Vol. 26, No. 1, 1988, p. 85;
    Advanced Problem H-430, Vol. 29, No. 1, 1991, p. 92;
    Elementary Problem B-741, Vol. 31, No. 2, 1993, p. 182.

4

Paper-Folding, Polyhedra-Building, and Number Theory

4.1 INTRODUCTION: FORGING THE LINK BETWEEN GEOMETRIC PRACTICE AND MATHEMATICAL THEORY

In this chapter we carry the paper-folding procedures and the mathematics of paper-folding further than we did in [2]. However, in order to make this account as self-contained as possible, we will recall, in Section 2, the systematic folding procedures from Chapter 4 of [2] that enabled us to approximate, to any degree of accuracy desired, any regular convex N-gon.[1] We will see, from the examples, that the process also enables us to fold certain regular star $\{\frac{b}{a}\}$-gons,[2] some of which are shown in Figure 1. For brevity we will refer to the approximations we obtain for both the regular convex N-gons and the regular star $\{\frac{b}{a}\}$-gons (when $a \geq 2$) as ***quasi-regular polygons***. In most cases the context will make it unnecessary to state whether or not they are genuine convex polygons. Sometimes we

---

[1] It is not uncommon for people to adopt special conventionally-permitted paper-folding procedures that produce exact constructions for certain families of regular polygons (see, for example, [13]).

[2] We will give a more precise definition of these star polygons in Section 2. Note that, when we speak of a $\{\frac{b}{a}\}$-gon, we assume that a, b are coprime; but, if a, b emerge from a *calculation* (see (3), (4) of Section 2), they may not, at that stage, be coprime.

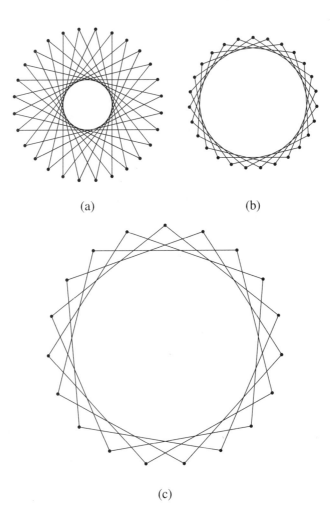

(a)   (b)

(c)

FIGURE 1 Some star $\{\frac{b}{a}\}$-gons. (a) $a = 28, b = 11$. (b) $a = 27, b = 5$. (c) $a = 19, b = 4$.

refer to a *star b-gon* to mean a star $\{\frac{b}{a}\}$-gon for some a prime to b and satisfying $a < \frac{b}{2}$.

In the process of recalling the folding procedures we will also describe again how to construct a ***symbol*** that enables one to read off the folding instructions for constructing any quasi-regular N-gon, where N is odd and ≥ 3. Surprisingly, the construction of this symbol led us, in [2], to the Quasi-Order Theorem in base 2 in number theory (the "2" came naturally from the paper-folding, since the folding procedure always involved *bisecting* angles). We will restate that remarkable theorem in Section 2 (the proof begins on page 130 of [2]) and illustrate it with some examples.

In Section 3 we describe precisely how to make secondary fold lines, to augment the primary fold lines described in Section 2, so that we can construct *any* quasi-regular polygon.

Section 4 gives instructions for how to use the folded strips of paper that produce 3-, 4-, 5-gons to build certain polyhedra. Many of these models are very striking to behold — but they have other unusual characteristics. Some of them come apart into straight strips, thus making them easy to store, others collapse in unexpected ways, and, as we show in Chapter 8, some are also connected with beautiful mathematical ideas; and we believe that this connection is no accident. Thus Section 4 is not itself mathematical in nature, but it can stimulate and enliven some fine mathematics.

In Section 5 we generalize the Quasi-Order Theorem to a general base *t*. In doing this we then are forced to give up the interpretation of folding paper, but the analogy is clear, and the result is truly remarkable. The proof of the general theorem is scarcely more difficult than that of the Quasi-Order Theorem in base 2, but the statement of the generalization is not at all obvious. If you try generalizing the Quasi-Order Theorem of Section 2 before reading Section 5, we think you will see what we mean.

● ● ● **BREAK 1**

Why is it no restriction on the notion of a star $\{\frac{b}{a}\}$-gon to insist that $a < \frac{b}{2}$? [Hint: What would be the difference between the star $\{\frac{5}{2}\}$-gon and the star $\{\frac{5}{3}\}$-gon?]

4.2 WHAT CAN BE DONE WITHOUT EUCLIDEAN TOOLS

We begin by recalling how the question of whether or not, for a given N, it is possible to construct a regular N-gon using Euclidean tools (straight edge and compass) has fascinated people since the time of the ancient Greeks. In fact, Gauss (1777–1855) completely settled the question by proving that a Euclidean construction of a regular N-gon is possible *if and only if* the number of sides N is of the form $N = 2^c \prod \rho_i$, where the numbers ρ_i are distinct Fermat primes — that is, primes of the form $F_n = 2^{2^n} + 1$. Now, since F_n is only known to be prime for

$$F_0 = 3, \quad F_1 = 5, \quad F_2 = 17, \quad F_3 = 257, \quad F_4 = 65537,$$

it is clear that a Euclidean construction of a regular N-gon is known to exist for very few values of N; and even for these N we do not know, at the time of writing, an explicit construction in all cases.

Despite this restrictive result we still would like, somehow, to construct *all* regular polygons. Our approach (as in [2]) is to modify the question so that, instead of asking for an exact construction,[3] we ask:

> *For which $N \geq 3$ is it possible, systematically and explicitly, to construct quasi-regular (convex) N-gons by paper-folding?*

Surprisingly, as we will show, the answer to this question is: *all $N \geq 3$*. Furthermore, in showing precisely how this is done, we receive a bonus, that is, we will also be able to construct all possible quasi-regular $\{\frac{b}{a}\}$-gons.

Let us now begin by recalling the precise and fundamental folding procedure, involving a straight strip of paper with parallel edges. We suggest that you will find it useful to have a long strip of paper handy. Adding-machine tape or ordinary unreinforced gummed tape work well.

Assume that we have a straight strip of paper that has certain vertices marked on its top and bottom edges, at equally spaced intervals, and that also has *creases* or *folds* along straight lines emanating from the vertices at the top edge of the strip. Further assume that the creases at those vertices labeled A_{nk}, $n = 0, 1, 2, \cdots$ (see Figure 2), which are on the top edge, form identical angles of $\frac{a\pi}{b}$ with the top edge, with an identical angle of $\frac{a\pi}{b}$ between the crease along the lines $A_{nk}A_{nk+2}$ and the crease along $A_{nk}A_{nk+1}$ (as shown in Figure 2(a)). If we fold this strip on $A_{nk}A_{nk+2}$, as shown in Figure 2(b), and then twist the tape so that it folds on $A_{nk}A_{nk+1}$, as shown in Figure 2(c), the direction of the *top edge* of the tape will be rotated through an angle of $2(\frac{a\pi}{b})$. We call this process of **f**olding **a**nd **t**wisting the **FAT**-algorithm (see any of [4, 5, 6, 7, 8, 11]).

Now consider the A_{nk} along the top of the tape, with k fixed and n varying. If the FAT-algorithm is performed on a sequence of angles, each of measure $\frac{a\pi}{b}$, at the vertices given by $n = 0, 1, 2, \cdots, b - 1$, then the top of the tape will have turned through an angle of $2a\pi$. Thus the vertex A_{bk} will come into coincidence with A_0; and the top edge of the tape will have visited every ath vertex of a bounding regular convex b-gon, thus creating a quasi-regular $\{\frac{b}{a}\}$-gon. As an example, see Figure 6(c), where $a = 2$ and $b = 7$. (In order to fit with our usage of "N-gon" we make a slight adaptation of the Coxeter notation for star polygons (see [1]), so that

---

[3]Of course, in many cases, such as when $N = 2^c$ (with $c \geq 2$), we can give exact constructions.

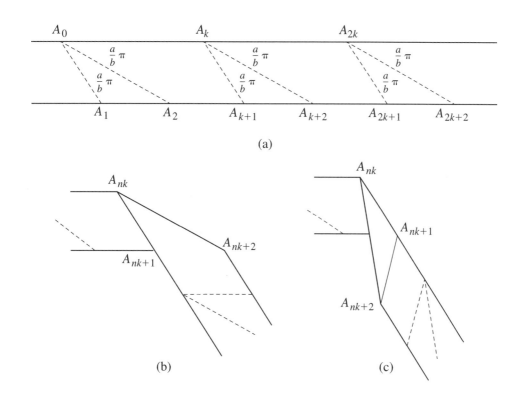

FIGURE 2

when we refer to a ***quasi-regular*** $\{\frac{b}{a}\}$***-gon*** we mean a connected sequence of edges that visits every ath vertex of a quasi-regular b-gon. Thus our N-gon is the special star $\{\frac{N}{1}\}$-gon. When labeling a convex polygon this way we may well use a lower case letter instead of N.)

Figure 3 illustrates how a suitably creased strip of paper may be folded by the FAT-algorithm to produce a quasi-regular p-gon, (or $\{\frac{p}{1}\}$-gon). In Figure 3 we have written V_k instead of A_{nk}, since it is more natural in this particular context.

Let us now illustrate how the FAT-algorithm may be used to fold a regular convex 8-gon. Figure 4(a) shows a straight strip of paper on which the dotted lines indicate certain special exact crease lines. In fact, these crease lines occur at equally spaced intervals along the top of the tape, so that the angles occurring at the top of each vertical line are (from left

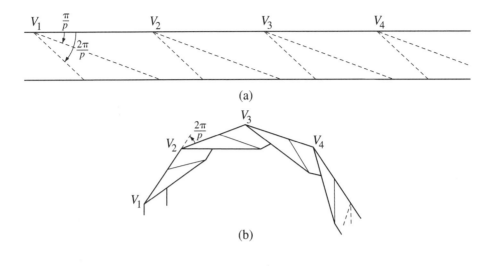

(a)

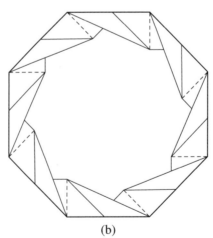

(b)

FIGURE 3

(a)

(b)

FIGURE 4

to right) $\frac{\pi}{2}$, $\frac{\pi}{4}$, $\frac{\pi}{8}$, $\frac{\pi}{8}$. Figuring out how to fold a strip of tape to obtain this arrangement of crease lines is very unlikely to cause the reader any difficulty, but complete instructions are given in [8]. Our immediate interest is focused on the observation that this tape has, at equally spaced intervals along the top edge, adjacent angles each measuring $\frac{\pi}{8}$, and we can therefore execute the FAT-algorithm at 8 consecutive vertices along the top of the tape to produce an exact regular convex 8-gon, as shown in Figure 4(b). (Of course, in constructing the model, one would cut the tape on the first vertical line and glue a section at the end to the beginning so that the model would form a closed polygon.)

Notice that the tape shown in Figure 4(a) also has suitable crease lines that make it possible to use the FAT-algorithm to fold a regular convex 4-gon. We leave this as an exercise for the reader and turn to a more challenging construction, the regular convex 7-gon.

Now, since the 7-gon is the first regular polygon that we encounter for which there does not exist a Euclidean construction, we are faced with a real difficulty in creating a crease line making an angle of $\frac{\pi}{7}$ with the top edge of the tape. We proceed by adopting a general policy we call our *optimistic strategy*. Assume that we *can* crease an angle of $\frac{2\pi}{7}$ (certainly we can come close) as shown in Figure 5(a). Given that we have the angle $\frac{2\pi}{7}$, it is then a trivial matter to fold the top edge of the strip DOWN to bisect this angle, producing two adjacent angles of $\frac{\pi}{7}$ at the top edge as shown in Figure 5(b). (We say that $\frac{\pi}{7}$ is the *putative* angle on this tape.) Then, since we are content with this arrangement, we go to the bottom of the tape, where we observe that the angle to the right of the last crease line is $\frac{6\pi}{7}$ — and we decide, as paper folders, that we will always avoid leaving even multiples of π in the numerator of any angle next to the edge of the tape, so we bisect this angle of $\frac{6\pi}{7}$, by bringing the bottom edge of the tape UP to coincide with the last crease line and creating the new crease line sloping up shown in Figure 5(c). We settle for this (because we are content with an odd multiple of π in the numerator) and go to the top of the tape, where we observe that the angle to the right of the last crease line is $\frac{4\pi}{7}$ — and, since we have decided against leaving an even multiple of π in any angle next to an edge of the tape, we are forced to bisect this angle twice, each time bringing the top edge of the tape DOWN to coincide with the last crease line, obtaining the arrangement of crease lines shown in Figure 5(d). But now we notice that something miraculous has occurred! If we had really started with an angle of exactly $\frac{2\pi}{7}$, and if we now continue introducing crease lines by repeatedly folding the tape DOWN TWICE at the top and UP ONCE at the bottom, we get precisely what

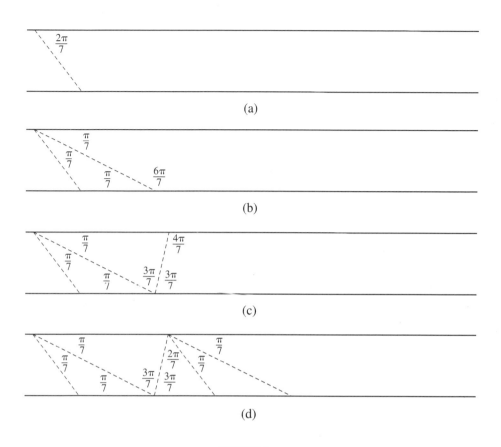

FIGURE 5

we want, namely, pairs of adjacent angles, measuring $\frac{\pi}{7}$, at equally spaced intervals along the top edge of the tape. Let us call this folding procedure the D^2U^1-*folding procedure* (or, more simply — and especially when we are concerned merely with the related number theory — the **(2, 1)**-*folding procedure*) and call the strip of creased paper it produces D^2U^1-*tape* (or, again more simply, **(2, 1)**-*tape*). The crease lines on this tape are called the *primary crease lines*.

● ● ● **BREAK 2**

(1) We suggest that before reading further you get a piece of paper and fold an acute angle which you call an approximation to $\frac{2\pi}{7}$. Then fold about 40 triangles using the D^2U^1-folding procedure

as shown in Figures 5 and 6(a) and described above, throw away the first 10 triangles, and see if you can tell that the first angle you get between the top edge of the tape and the adjacent crease line is *not* $\frac{\pi}{7}$. Then try to construct the FAT 7-gon shown in Figure 6(b). You may then *believe* that the D^2U^1-folding procedure produces tape on which the smallest angle does approach $\frac{\pi}{7}$, in fact rather rapidly.

(2) Try executing the FAT-algorithm at every other vertex along the top of this tape to produce a quasi-regular $\{\frac{7}{2}\}$-gon. (Hint: Look at Figure 6(c).)

How do we *prove* that this evident convergence actually takes place? A very direct approach is to admit that the first angle folded down from the top of the tape in Figure 5(a) might not have been precisely $\frac{2\pi}{7}$. Then the bisection forming the next crease would make the two acute angles nearest the top edge in Figure 5(b) only approximately $\frac{\pi}{7}$; let us call them $\frac{\pi}{7} + \epsilon$ (where the error ϵ may be either positive or negative). Consequently, the angle to the right of this crease, at the bottom of the tape, would measure $\frac{6\pi}{7} - \epsilon$. When this angle is bisected, by folding up, the resulting acute angles nearest the bottom of the tape, labeled $\frac{3\pi}{7}$ in Figure 5(c), would in fact measure $\frac{3\pi}{7} - \frac{\epsilon}{2}$, forcing the angle to the right of this crease line at the top of the tape to have measure $\frac{4\pi}{7} + \frac{\epsilon}{2}$. When this last angle is bisected twice by folding the tape down, the two acute angles nearest the top edge of the tape will measure $\frac{\pi}{7} + \frac{\epsilon}{2^3}$. This makes it clear that every time we repeat a D^2U^1-folding on the tape the error is reduced by a factor of 2^3.

We see that our ***optimistic strategy*** has paid off — by blandly *assuming* we have an angle of $\frac{\pi}{7}$ at the top of the tape to begin with, and folding accordingly, we *get what we want* — successive angles at the top of the tape that, as we fold, rapidly get closer and closer to $\frac{\pi}{7}$, whatever angle we had, in fact, started with!

In practice, the approximations we obtain by folding paper are quite as accurate as the *real world* constructions with a straight edge and compass — for the latter are only perfect in the mind. In both cases the real world result is a function of human skill, but our procedure, unlike the Euclidean procedure, is very forgiving in that it tends to reduce the effects of human error — and, for many people (even the not so young), it is far easier to bisect an angle by folding paper than it is with a straight edge and compass.

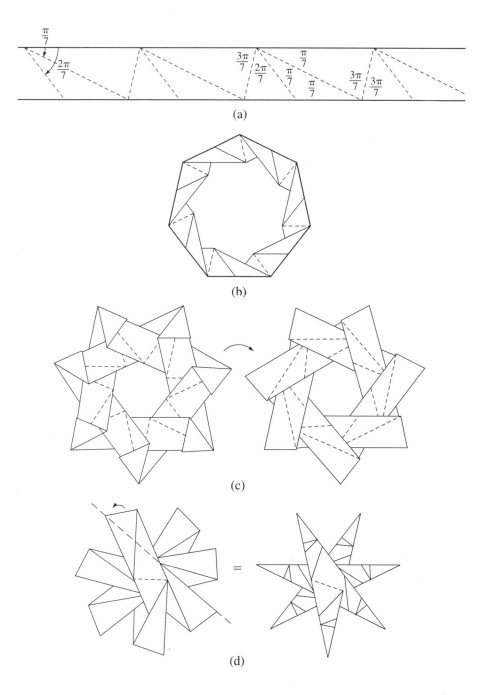

(a)

(b)

(c)

(d)

FIGURE 6

Figures 6(c), 6(d) show the regular $\{\frac{7}{2}\}$- and $\{\frac{7}{3}\}$-gons that are produced from the D^2U^1-tape by executing the FAT-algorithm on the crease lines that make angles of $\frac{2\pi}{7}$ and $\frac{3\pi}{7}$, respectively, with an edge of the tape (if the angle needed is at the bottom of the tape, as with $\frac{3\pi}{7}$, simply turn the tape over so that the required angle appears on the top). In Figures 6(c), 6(d) the FAT-algorithm was executed on every other suitable vertex along the edge of the tape so that, in (c), the resulting figure, or its flipped version, could be woven together in a more symmetric way and, in (d), the excess could be folded neatly around the points.

It is now natural to ask:

1. Can we use the same general approach used for folding a convex 7-gon to fold a convex N-gon with N odd, at least for certain specified values of N? If so, can we always prove that the actual angles on the tape really converge to the putative angle we originally sought?
2. Do we always get a quasi-regular $\{\frac{b}{a}\}$-gon with any general folding procedure, perhaps with other periods, such as those represented by

$$D^3U^3, \quad D^4U^2, \quad \text{or} \quad D^3U^1D^1U^3D^1U^1?$$

How does the folding procedure determine $\frac{b}{a}$?

(The **period** is determined by the repeat of the *exponents*, so these examples have periods 1, 2, and 3, respectively.)

The answer to (1) is *yes*, and we will soon show you an algorithm for determining the folding procedure that produces tape from which you can construct *any* given quasi-regular $\{\frac{b}{a}\}$-gon, if a, b are odd with $a < \frac{b}{2}$. The complete answer to (2) appears in [2], but here we will simply note that an iterative folding procedure of this type will always produce one and only one quasi-regular $\{\frac{b}{a}\}$-gon (see page 135 of [2]).

Let us now look at the general 1-*period* folding procedure D^nU^n. A typical portion of the tape would appear as illustrated in Figure 7(a).

It turns out that the smallest angle u_k at the top, or bottom, of this tape approaches $\frac{\pi}{2^n+1}$; that is,

$$u_k \longrightarrow \frac{\pi}{2^n + 1} \quad \text{as} \quad k \longrightarrow \infty \qquad (1)$$

A proof of (1) similar to the one provided above for the tape whose smallest angle approached $\frac{\pi}{7}$ may be given. In fact, we can see that, if the original fold down (supposedly making an angle of $\frac{2\pi}{2^n+1}$ with the top of the tape) were such that it produced an angle that differed from the

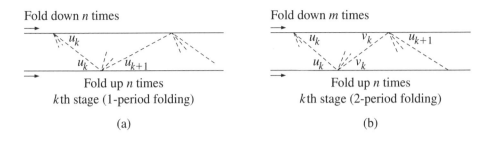

Fold down n times

Fold up n times

kth stage (1-period folding)

(a)

Fold down m times

Fold up n times

kth stage (2-period folding)

(b)

FIGURE 7

putative angle $\frac{\pi}{2^n+1}$ by an error ϵ_0, then the error ϵ_k at the kth stage of the $D^n U^n$-folding procedure would be given by

$$|\epsilon_k| = \frac{|\epsilon_0|}{2^{nk}} \qquad (2)$$

Hence we see that the $D^n U^n$-folding procedure produces tape from which we may construct quasi-regular $(2^n + 1)$-gons — and, of course, these include those N-gons for which N is a Fermat number, prime or not. We would like to believe that the ancient Greeks and Gauss would have appreciated the fact that, when $n = 1, 2, 4, 8,$ and 16, the $D^n U^n$-folding procedure produces tape from which we can obtain, by means of the FAT-algorithm, a quasi-regular 3-, 5-, 17-, 257-, and 65537-gon, respectively. What's more, if $n = 3$, we approximate the regular 9-gon, whose non-constructibility by Euclidean tools is very closely related to the non-trisectibility of an arbitrary angle.

The case $N = 2^n + 1$ is atypical, since we may construct $(2^n + 1)$-gons from our folded tape by special methods (not involving the FAT-algorithm), in which, however, the top edge does not describe the polygon, as it does in the FAT-algorithm. Figure 8 shows how the $D^2 U^2$-tape shown in part (a) may be folded along just the short lines of the creased tape to form the *outline* of a quasi-regular pentagon shown in (b), and along just the long lines of the creased tape to form the *outline* of the slightly larger quasi-regular pentagon shown in (c); and, finally, we show in (d) the quasi-regular pentagon formed by an edge of the tape when the FAT-algorithm is executed.

• • • **BREAK 3**

Fold a length of $D^n U^n$-tape, for various values of n, and *experiment* with the folded tape to see how many differently-sized regular

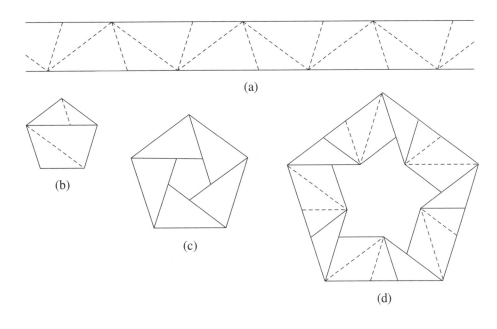

FIGURE 8

$(2^n + 1)$-gons you can create from each tape. You will need the D^1U^1- and the D^2U^2-tape in the section on building models, so be sure to include $n = 1$ and 2 in your experiments. Figure 9 shows three possibilities for constructing a quasi-regular 9-gon from D^3U^3-tape, without using the FAT algorithm. (The shaded portion of the figure indicates that the reverse side of the tape is visible.)

We next demonstrate how we construct quasi-regular polygons with 2^cN sides, N odd, if we already know how to construct quasi-regular N-gons. If, for example, we wished to construct a quasi-regular 10-gon, then we take the D^2U^2-tape (which, as you may recall, produced FAT 5-gons) and introduce a **secondary crease line** by bisecting each of the angles of $\frac{\pi}{5}$ next to the top (or bottom) edge of the tape. The FAT-algorithm may be used on the resulting tape to produce the quasi-regular convex FAT 10-gon, as illustrated in Figure 10. It should now be clear how to construct a quasi-regular 20-gon, 40-gon, 80-gon, $\cdots$.

This argument shows that we only need construct quasi-regular N-gons for N odd in order to be able to construct quasi-regular N-gons for any N.

Now we turn to the general **2-period** folding procedure, D^mU^n, which we may abbreviate to (m, n). (Recall that the tape that produced the quasi-

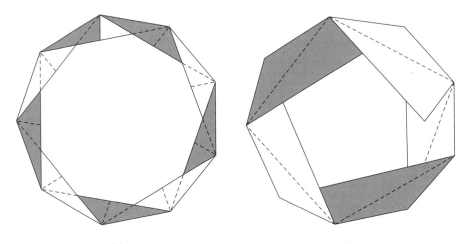

(a)   (b)

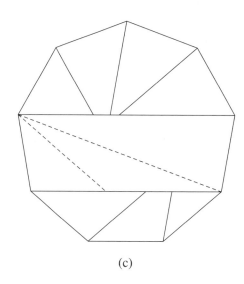

(c)

FIGURE 9   (a) A *long*-line 9-gon. (b) A *medium*-line 9-gon. (c) A *short*-line 9-gon.

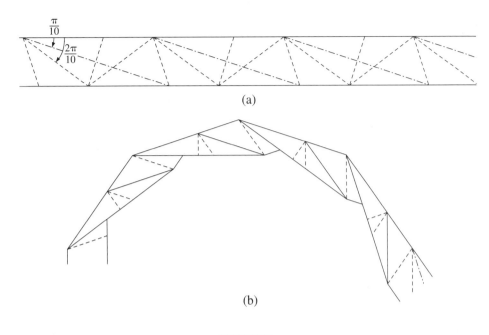

$\dfrac{\pi}{10}$

$\dfrac{2\pi}{10}$

(a)

(b)

FIGURE 10

regular 7-gon was a 2-period tape employing the $(2, 1)$ procedure.) A typical portion of the 2-period tape, in the general case, may be illustrated as shown in Figure 7(b). If the folding procedure had been started with an arbitrary angle u_0 at the top of the tape, and continued producing angles $u_1, u_2, \cdots$ at the top and $v_0, v_1, \cdots$ at the bottom, we would have, at the kth stage,

$$u_k + 2^n v_k = \pi,$$
$$v_k + 2^m u_{k+1} = \pi,$$

and it is shown in [2] that then

$$u_k \longrightarrow \frac{2^n - 1}{2^{m+n} - 1}\pi \quad \text{as} \quad k \to \infty \tag{3}$$

so that $\frac{2^n-1}{2^{m+n}-1}\pi$ is the putative angle. Thus the FAT-algorithm will produce, from this tape, a star $\{\frac{b}{a}\}$-gon, where the fraction $\frac{b}{a}$ may turn out not to be reduced (for example, when $n = 2$, $m = 4$), with $b = 2^{m+n} - 1$, $a = 2^n - 1$. By symmetry we infer that

$$v_k \longrightarrow \frac{2^m - 1}{2^{m+n} - 1}\pi \quad \text{as} \quad k \to \infty \tag{4}$$

Furthermore, if we assume an initial error of ϵ_0, then it can be shown (see [2]) that the error at the kth stage (when the folding $D^m U^n$ has been done exactly k times) will be given by[4]

$$\epsilon_k = \frac{\epsilon_0}{2^{(m+n)k}} \tag{5}$$

Hence, we see that in the case of our $D^2 U^1$-folding (Figures 5, 6(a)) any initial error ϵ_0 is, as we already saw from our other argument, reduced by a factor 8 between consecutive stages. It should now be clear why we advised throwing away the first part of the tape — but, likewise, it should also be clear that it is never necessary to throw away very much of the tape. In practice, convergence is very rapid indeed, and if one made it a rule of thumb to always throw away the first 20 crease lines on the tape for any iterative folding procedure, it would turn out to be a very conservative rule.

We see that, however wonderful these results may be, they haven't completely solved our problem. For example, as we will have you show in the next break, we would be unable to fold a quasi-regular 11-gon with either the 1- or 2-period folding procedure. So the question remains: ***how do we know which sequence of folds to make in order to produce a particular quasi-regular polygon with the FAT-algorithm?***

● ● ●   **BREAK 4**

(1) Show that $\frac{2^{m+n}-1}{2^n-1}$ will be an integer if and only if $n|m$. (Hint: Write the top and bottom of $\frac{2^{m+n}-1}{2^n-1}$ in base 2 and carry out the division.. Try some examples and pay particular attention to the form the *remainder* takes.)

(2) Show that, even if $n|m$, the number $\frac{2^{m+n}-1}{2^n-1}$ is never equal to 11. (Hint: In the same division, pay special attention to the *quotient*.)

We now show, with a particular but not special case, how to determine the folding instructions for producing tape from which we can construct a quasi-regular $\{\frac{b}{a}\}$-gon, with a, b odd and $a < \frac{b}{2}$.

---

[4]The discrepancy between formulae (2) and (5) is due to the fact that the $D^n U^n$ folding procedure is really a period 1 procedure.

Thus, suppose we want to construct a quasi-regular $\{\frac{11}{3}\}$-gon. Then, of course, $b = 11$, $a = 3$, and we proceed precisely as we did when we wished to construct the regular convex 7-gon; that is, we adopt our *optimistic strategy* which, as you recall, means that we *assume* that we've got what we want, and, as we will show, we then actually *get* an arbitrarily good approximation to what we want! This time we assume that we can fold the desired putative angle of $\frac{3\pi}{11}$ at A_0 (see Figure 11(a)), and we adhere to the same principles that we used in constructing the quasi-regular 7-gon, namely, we adopt the following rules.

1. Each new crease line goes in the forward (left to right) direction along the strip of paper.
2. Each new crease line always *bisects* the angle between the last crease line and the edge of the tape from which it emanates.
3. The bisection of angles at any vertex continues until a crease line produces an angle of the form $\frac{a'\pi}{b}$ where a' is an *odd* number; then the folding stops at that vertex and commences at the intersection point of the last crease line with the other edge of the tape.

Once again the *optimistic strategy* works; and following this procedure results in tape whose angles converge to those shown in Figure 11(b). We could denote this folding procedure by $D^1 U^3 D^1 U^1 D^3 U^1$, interpreted in the obvious way on the tape — that is, the first exponent "1" refers to the one bisection (producing a line in a downward direction) at the vertices A_{6n} (for $n = 0, 1, 2, \ldots$) on the top of the tape; similarly, the "3" refers to the 3

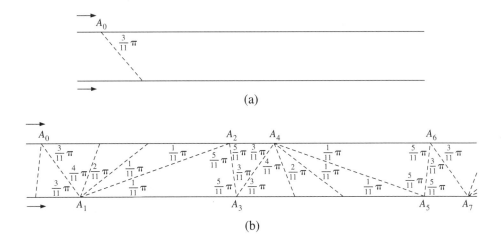

(a)

(b)

FIGURE 11 (Note that the indexing of the vertices is not the same as that in Figure 3.)

bisections (producing creases in an upward direction) made at the bottom of the tape through the vertices A_{6n+1}; etc. However, since the folding procedure is *duplicated* halfway through, we can abbreviate the notation and write simply $\{1, 3, 1\}$, with the understanding that we alternately fold from the top and bottom of the tape as described, with the *number* of bisections at each vertex running, in order, through the values $1, 3, 1, \cdots$. We call this a ***primary folding procedure of period 3*** or a ***3-period folding***.

To prove the convergence we can use an error-correction type of proof like that given earlier in this section for the 7-gon. We leave the details to the reader, and explore here what we can do with this $(1, 3, 1)$-tape. First, note that, starting with the putative angle $\frac{3\pi}{11}$ at the top of the tape, we produce a putative angle of $\frac{\pi}{11}$ at the bottom of the tape, then a putative angle of $\frac{5\pi}{11}$ at the top of the tape, then a putative angle of $\frac{3\pi}{11}$ at the *bottom* of the tape, and so on. A careful inspection of this tape shows that we could use the FAT algorithm on it to fold quasi-regular $\left\{\frac{11}{a}\right\}$-gons, when $a = 1, 2, 3, 4, 5$. To put the result in a form that suggests the generalization, we may say that if there are crease lines enabling us to fold a star $\{\frac{11}{a}\}$-gon, there will be crease lines enabling us to fold star $\{\frac{11}{2^k a}\}$-gons, where $k \geq 0$ takes any value such that $2^{k+1}a < 11$. These features, described for $b = 11$, would be found with any odd number b. However, this tape has a special

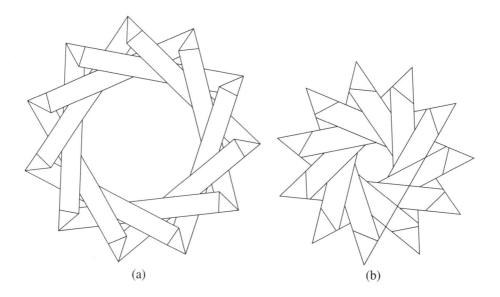

(a)   (b)

FIGURE 12

symmetry as a consequence of its *odd* period; namely, if it is "flipped" about the horizontal line halfway between its parallel edges, the result is a *translate* of the original tape. As a practical matter this special symmetry of the tape means that we can use either the top edge or the bottom edge of the tape to construct our polygons. On tapes with an *even* period the top edge and the bottom edge of the tape are not translates of each other (under the horizontal flip), which simply means that care must be taken in choosing the edge of the tape used to construct a specific polygon. Figures 12(a, b) show the completed $\{\frac{11}{3}\}$-, $\{\frac{11}{4}\}$-gons, respectively.

Now, to set the scene for the number theory of Section 5, and to enable us to systematically determine the folding procedure for any given a and b, let us look at the patterns in the *arithmetic* of the computations when $a = 3$ and $b = 11$. Referring to Figure 10(b) we observe that

the smallest angle to the right of A_n, where	is of the form $\frac{a}{11}\pi$, where	and the number of bisections at the *next* vertex[5]
$n = 0$	$a = 3$	$= 3$
1	1	1
2	5	1
3	3	3
4	1	1
5	5	1

We could write this in shorthand form as follows:

$$(b =)11 \left| \begin{array}{ccc} (a =)3 & 1 & 5 \\ 3 & 1 & 1 \end{array} \right. \tag{6}$$

Observe that, had we started with the putative angle of $\frac{\pi}{11}$, then the *symbol* (6) would have taken the form

$$(b =)11 \left| \begin{array}{ccc} (a =)1 & 5 & 3 \\ 1 & 1 & 3 \end{array} \right. \tag{6'}$$

In fact, it should be clear that we can *start anywhere* (with $a = 1, 3,$ or 5), and the resulting symbol, analogous to (6), will be obtained by cyclic

---

[5]Notice that, referring to Figure 10(b), to obtain an angle of $\frac{3\pi}{11}$ at $A_0, A_6, A_{12}, \cdots$, the folding instructions would more precisely be $U^3 D^1 U^1 D^3 U^1 D^1 \cdots$. But we don't have to worry about this distinction.

permutation of the matrix component of the symbol, placing our choice of a in the first position along the top row.

In general, suppose we wish to fold a $\{\frac{b}{a}\}$-gon, with b, a odd and $a < \frac{b}{2}$. Then we may construct a symbol[6] as follows. Let us write

$$
b \begin{vmatrix} a_1 & a_2 & \cdots & a_r \\ k_1 & k_2 & \cdots & k_r \end{vmatrix}
\tag{7}
$$

where b, a_i $(a_1 = a)$ are odd, $a_i < \frac{b}{2}$, and

$$
b - a_i = 2^{k_i} a_{i+1}, \quad i = 1, 2, \cdots, r, \quad a_{r+1} = a_1
\tag{8}
$$

We proved in Chapter 4 of [2], that, given any two odd numbers a and b, with $a < \frac{b}{2}$, there is always a completely determined unique symbol (7) with $a_1 = a$. At this stage, we do not assume that $\gcd(b, a) = 1$, but we have assumed that the list $a_1, a_2, \cdots, a_r$ is without repeats. Indeed, if $\gcd(b, a) = 1$, we say that the symbol (7) is *reduced*, and, if there are no repeats among the a_i's, we say that the symbol (7) is *contracted*. (It is, of course, theoretically possible to consider symbols (7) in which repetitions among the a_i are allowed.) We regard (7) as encoding the general folding procedure to which we have referred.

Example 1 If we wish to fold a 31-gon we may start with $b = 31, a = 1$ and construct the symbol

$$
(b =)31 \begin{vmatrix} (a =)1 & 15 \\ 1 & 4 \end{vmatrix}
$$

which tells us that folding $D^1 U^4$ will produce tape (usually called $(1, 4)$-tape) that can be used to construct a FAT 31-gon. In fact, this tape can also be used to construct FAT

$$
\left\{ \frac{31}{2} \right\} -, \quad \left\{ \frac{31}{4} \right\} -, \quad \text{and} \quad \left\{ \frac{31}{8} \right\} \text{-gons.}
$$

However, if we wish to fold a $\{\frac{31}{3}\}$-gon, we start with $b = 31, a = 3$ and construct the symbol

$$
(b =)31 \begin{vmatrix} (a =)3 & 7 \\ 2 & 3 \end{vmatrix}
$$

[6]More exactly, a 2-symbol. Later on, we introduce a more general t-symbol, $t \geq 2$.

which tells us to fold D^2U^3 — or, more simply, to use the $(2, 3)$-folding procedure — to produce $(2, 3)$-tape from which we can fold the FAT $\{\frac{31}{3}\}$-gon. Again, we get more than we initially sought, since we can also use the $(2, 3)$-tape to construct FAT

$$\left\{\frac{31}{6}\right\}\text{-,} \quad \left\{\frac{31}{12}\right\}\text{-,} \quad \left\{\frac{31}{7}\right\}\text{-,} \quad \text{and} \quad \left\{\frac{31}{14}\right\}\text{-gons.}$$

However, we don't have a folding procedure that produces the $\{\frac{31}{5}\}$-gon. Thus we construct another symbol, this time with $b = 31$, $a = 5$. We get

$$(b =)31 \ \left| \ (a =)5 \quad 13 \quad 9 \quad 11 \atop \quad\quad 1 \quad\ \ 1 \quad 1 \quad 2 \right|$$

which tells us to fold $D^1U^1D^1U^2$ — or, more simply, to use the 4-period $(1, 1, 1, 2)$-folding procedure — to produce $(1, 1, 1, 2)$-tape from which we can fold the FAT $\{\frac{31}{5}\}$-gon. Once again, we get more than we asked for; we can also use the $(1, 1, 1, 2)$-tape to construct FAT

$$\left\{\frac{31}{10}\right\}\text{-,} \quad \left\{\frac{31}{13}\right\}\text{-,} \quad \left\{\frac{31}{9}\right\}\text{-,} \quad \text{and} \quad \left\{\frac{31}{11}\right\}\text{-gons.}$$

We can combine all the possible symbols for $b = 31$ into one *complete* symbol, adopting the notation

$$31 \ \left| \begin{array}{cc|cc|cccc} 1 & 15 & 3 & 7 & 5 & 13 & 9 & 11 \\ 1 & 4 & 2 & 3 & 1 & 1 & 1 & 2 \end{array} \right| \tag{9}$$

Notice in (9) that the total amount of folding would be the same to produce any quasi-regular (convex or star) 31-gon. Since it is very difficult to bisect an angle 4 times, you may prefer to use the second or third parts of this symbol to produce the tape. Even if you really want a convex 31-gon it may be easier, in practice, to produce the star polygon first and then use the vertices of that polygon to determine the convex polygon.

••• **BREAK 5**

(1) Show that the tape folded in accordance with the folding instruction, or symbol,

$$b \ \left| \begin{array}{cccc} a_1 & a_2 & \cdots & a_r \\ k_1 & k_2 & \cdots & k_r \end{array} \right.$$

contains fold lines allowing us, by means of the FAT-algorithm, to fold any quasi-regular $\{\frac{b}{2^n a_i}\}$-gon, where $2^{n+1} a_i < b$.

Show that the complete symbol for $b = 31$ gives us, in this sense, folding instructions for folding all possible quasi-regular star 31-gons. Give an argument for why this must happen for any odd b. (Note that we are only considering star $\{\frac{b}{a}\}$-gons where a is prime to b.)

(2) Check your understanding of the complete symbol by doing the calculations to fill in the blanks below for $b = 91$:

$$
91 \ \begin{vmatrix} 1 \ 45 \ 23 \ 17 \ 37 \ 27 & 3 \ 11 \ 5 \ 43 & 9 \ 41 \ 25 \ 33 \ 29 \ 31 \ 15 \ 19 \\ 1 \ 1 \ 2 \ 1 \ 1 \ 6 & 3 \ ? \ ? \ 4 & 1 \ ? \ ? \ ? \ ? \ ? \ ? \ ? \end{vmatrix}
$$

(Notice that no multiples of 7 or 13 appear in the top row. Why do you suppose this is so?)

(3) Calculate other complete symbols for odd b and look for patterns in them.

(4) Calculate the symbol with $b = 33, a = 9$. Compare with item (6). What part of this symbol tells you that the rational number $\frac{33}{9}$ is equal to $\frac{11}{3}$? (You should now see why it is pointless to allow unreduced symbols.)

(5) Guess what the values of k_i would be for the symbol with $b = 91$, $a = 13$. Construct the symbol to see if your guess was correct.

More formally, we can say that given positive odd integers b, a with $a < \frac{b}{2}$, there is always a unique contracted symbol (the proof is in [2])

$$
b \ \begin{vmatrix} a_1 & a_2 & \cdots & a_r \\ k_1 & k_2 & \cdots & k_r \end{vmatrix}, \qquad a_1 = a, \quad a_i \neq a_j \ \text{ if } \ i \neq j, \qquad (10)
$$

where each a_i is odd, $a_i < \frac{b}{2}$, and

$$
b - a_i = 2^{k_i} a_{i+1}, \quad i = 1, 2, \cdots, r, \quad a_{r+1} = a_1 \qquad (11)
$$

The proof involves fixing b and letting S be the set of positive odd numbers $a < \frac{b}{2}$. Given $a \in S$, then a' is defined by the rule

$$
b - a = 2^k a', \quad k \ maximal; \qquad (12)
$$

that is, we take as many factors of 2 as we can out of $b - a$. Then (12) describes a function $\Psi : S \longrightarrow S$ such that $\Psi(a) = a'$. In [2] we show that Ψ is a **permutation** of the finite set S.

The permutation Ψ has the important property

$$\gcd(b, a) = \gcd(b, a'). \tag{13}$$

For it is clear from (12) that if $d \mid b$ and $d \mid a'$, then $d \mid b$ and $d \mid a$. Conversely, if $d \mid b$ and $d \mid a$, then d is *odd* and $d \mid 2^k a'$, so $d \mid b$ and $d \mid a'$. Thus if a_1 in (10) is prime to b, so are $a_2, a_3, \cdots, a_r$, and we may, if we wish, confine our choices of a_i to those odd numbers such that $\gcd(a_i, b) = 1$; that is, we may confine ourselves to **reduced** symbols, now to be defined as those symbols in which *each a_i is prime to b*. Moreover, given any odd numbers b, a with $a < \frac{b}{2}$, we may construct the symbol (10) and then reduce the symbol by dividing b and each a_i by $\gcd(b, a)$; the bottom row will be unaffected.

4.3 CONSTRUCTING ALL QUASI-REGULAR POLYGONS

We have described procedures for folding any quasi-regular convex polygon and for folding any quasi-regular star $\{\frac{b}{a}\}$-gon if a, b are both odd with, of course, $a < \frac{b}{2}$. To complete our program we must show how to fold any quasi-regular star $\{\frac{b}{a}\}$-gon, where $a < \frac{b}{2}$ and (i) b is odd, a is even or (ii) b is even, a is odd (for we may, of course, assume a, b coprime).

The case where b is odd and a is even is quickly dealt with. Let $a = 2^k a'$ with a' odd, and suppose we have tape creased to fold a $\{\frac{b}{a'}\}$-gon. We claim that this tape already has a crease line making an angle $\frac{a\pi}{b}$ with the forward direction of the top of the tape, with another crease line making an angle $\frac{a\pi}{b}$ with it. For, if ℓ is minimal such that $2^\ell a' > \frac{b}{2}$, then $\ell \geq 1$ and the tape has a crease line making an angle of $\frac{2^\ell \pi a'}{b}$ with the forward direction of the top of the tape, appropriately bisected by crease lines ℓ times. (See Figure 13 for a typical example, with $\ell = 2$.) Now $2^\ell a' > \frac{b}{2}$, so $\ell > k$; thus the stated crease lines appear on the tape, and the FAT algorithm may be applied

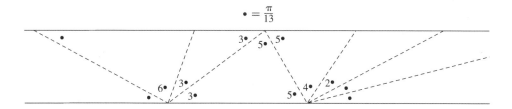

FIGURE 13 $a = 6, a' = 3, k = 1, b = 13$

using these two crease lines to fold a $\{\frac{b}{a}\}$-gon. No further crease lines are necessary.

The situation is very different, however, when b is even and a is odd — of course, we have already seen this on page 101 of Chapter 4 of [2] when $a = 1$. We now assume that $b = 2^k b'$, with b' odd. We write a in base 2 as

$$a = \epsilon_0 \epsilon_1 \cdots \epsilon_n, \quad \epsilon_0 = \epsilon_n = 1, \quad \epsilon_i = 0 \text{ or } 1, \quad 1 \leq i \leq n - 1. \quad (14)$$

We now describe the *initial configuration* of our tape. It will consist of two crease lines, the first making an angle of $\alpha_0 = \frac{\pi}{2^{k-n} b'}$ with the top of the tape, the second making an angle of $\sigma_0 = \frac{\pi}{2^{k-n} b'}$ with the first. To explain how we achieve this initial configuration, we must consider 3 cases.

> **Case 1:** $b' = 1$, so that $b = 2^k$. Then $2^n < a < 2^{k-1}$, so $n \leq k - 2$. We can then certainly achieve our initial configuration exactly.
>
> **Case 2:** $b' > 1$, $k > n$. Start with the b'-tape and create the initial configuration by introducing $(k - n)$ secondary crease lines by successively bisecting the top angle on the tape.
>
> **Case 3:** $b' > 1$, $k \leq n$. Now the b'-tape will itself already have the initial configuration on it — the argument is exactly like that above (when b is odd and a is even).

Thus we can suppose the initial configuration achieved. We are now going to define inductively the ***i*th proximand** α_i and the ***i*th support** σ_i, $0 \leq i \leq n$, and explain how they are achieved by folding the tape. We claim that on achieving the nth proximand we will have a crease line making an angle of $\frac{\pi a}{b}$ with the forward direction of the tape. Obviously, we can then duplicate this angle under this crease line and then apply the FAT algorithm to complete the construction of the $\{\frac{b}{a}\}$-gon.

Thus we suppose α_i, σ_i already defined and achieved, $0 \leq i < n$, with a crease line making an angle of α_i with the top of the tape and another crease line making an angle of σ_i with it; see Figure 14. Bisect σ_i by a new crease line into two (equal) angles λ_i, ρ_i, with λ_i to the left of ρ_i. If $\epsilon_{i+1} = 0$, define $\alpha_{i+1} = \alpha_i$, $\sigma_{i+1} = \rho_i$; if $\epsilon_{i+1} = 1$, define $\alpha_{i+1} = \alpha_i \cup \rho_i$, $\sigma_{i+1} = \lambda_i$.

We now prove our claim that $\alpha_n = \frac{\pi a}{b}$. In fact, we prove inductively that

$$\alpha_i = \frac{\pi a_i}{2^k b'}, \quad \sigma_i = \frac{\pi}{2^{k-n+i} b'}, \quad (15)$$

where

$$a_i = \epsilon_0 \epsilon_1 \cdots \epsilon_i 0 \cdots 0, \quad \text{with } (n - i) \text{ zeros.} \quad (16)$$

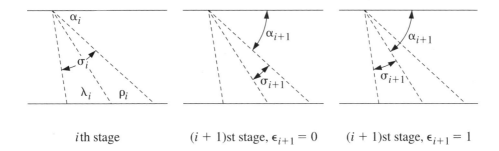

ith stage $(i + 1)$st stage, $\epsilon_{i+1} = 0$ $(i + 1)$st stage, $\epsilon_{i+1} = 1$

FIGURE 14

Then the equations (15) hold if $i = 0$. Given (15) for a fixed i, assume first that $\epsilon_{i+1} = 0$. Then $\alpha_{i+1} = \alpha_i = \frac{\pi a_{i+1}}{2^k b'}$, since $a_{i+1} = a_i$, and $\sigma_{i+1} = \frac{1}{2}\sigma_i = \frac{\pi}{2^{k-n+i+1}b'}$.

Assume instead that $\epsilon_{i+1} = 1$. Then

$$\alpha_{i+1} = \alpha_i + \frac{1}{2}\sigma_i = \frac{\pi a_i}{2^k b'} + \frac{\pi}{2^{k-n+i+1}b'} = \frac{\pi}{2^k b'}\left(a_i + 2^{n-i-1}\right)$$

$$= \frac{\pi}{2^k b'}\left(\epsilon_0\epsilon_1\cdots\epsilon_i 10\cdots 0\right) = \frac{\pi a_{i+1}}{2^k b'}.$$

Again $\sigma_{i+1} = \frac{1}{2}\sigma_i = \frac{\pi}{2^{k-n+i+1}b'}$. This establishes the inductive step and hence (15). Thus $\alpha_n = \frac{\pi a_n}{2^k b'} = \frac{\pi a}{b}$, and our construction has been vindicated.

Of course, what we have given above are algorithms for producing any quasi- regular star polygon by folding paper. Thus it is not to be expected that our recipes will give you the *simplest* procedures in all cases. Indeed, as you will see, they do not.

● ● ● **BREAK 6**

(1) Give an algorithm that works if $b = 2^k b'$, $a < \frac{b'}{2}$, b' odd, which is simpler than that given in the absence of the condition $a < \frac{b'}{2}$.

(2) Compare the algorithm in the text with the one you discovered in tackling problem (1) above, when $a = 1$.

4.4 HOW TO BUILD SOME POLYHEDRA (HANDS-ON ACTIVITIES)

In this section we give you explicit instructions for using the $D^1 U^1$-tape to construct regular pentagonal (and triangular) dipyramids, tetrahedra, octahedra, and icosahedra. We also show you how to make exact folds on the tape in order to be able to construct two different kinds of cube. Finally, we will show you how to use the D^2U^2-tape to construct two kinds of regular dodecahedron. All of these models may be taken apart and stored flat.

The models described in this section are referred to in Chapter 8, where some of the mathematics connected with their symmetries is discussed.

First we will tell you what you need. After you have assembled the materials you should choose which model you wish to make and then carefully read, and execute, the instructions for preparing the pattern pieces. Then we suggest that, after a rest, you read the assembly instructions.

What you will need

- A 75 ft (or more) roll of 2-in gummed mailing tape (or a wider and longer roll if you want larger models). The glue on the tape should be the type that needs to be moistened to become sticky. ***Caution:*** Don't try to use tape that is sticky to the touch when it is dry — you would find it very frustrating.
- Scissors
- Sponge (or washcloth)
- Shallow bowl
- Water
- Hand towel (or rag)
- Some books
- Colored paper of your choosing. Construction paper works well, but it may need to be cut into strips and glued together to get long enough pieces. Gift wrapping or butcher paper that comes in rolls are particularly easy to use for these models. The Sunday funnies also work.
- Bobby pins

General instructions for preparing pattern pieces

For each of the models described in this section you will need to glue the pattern piece (or pieces) onto colored paper. Of course, if a model involves more than one piece, the finished model will be more interesting if you use a different color for each piece. In each case, to accomplish the gluing, first prepare the pieces of paper onto which you plan to glue the *prepared* pattern piece. Make certain that each piece is long enough for the pattern piece and that it all lies on a flat surface.

Place a sponge (or washcloth) in a bowl. Add water to the bowl so that the top of the sponge is very moist indeed.[7] Moisten one end of the pattern piece by pressing it onto the sponge; then, holding that end (yes, it's messy!), pull the rest of the strip across the sponge. Make certain the entire strip gets wet and then place it on the colored paper. Use a hand towel (or rag) to wipe up the excess moisture while smoothing the tape into contact with the colored paper.

Put some books on top of the pieces so that they will dry flat. When the tape is dry, cut out the pattern pieces, *trimming off a small amount of the gummed tape* (about $\frac{1}{16}$ of an inch or 1.5 mm will do) from the edge as you do so (this serves to make the model look neater and, more importantly, allows for the increased thickness produced by gluing the strip to another piece of paper). Refold the piece *firmly* on all the fold lines that are going to be fold lines on the finished model. This should be done so that the raised ridges, called *mountain folds*, are *on the colored side* of the pattern piece. You will now be ready to construct your model.

We now describe the specific details for each model:

Pentagonal Dipyramid constructed from one strip

Begin by folding the gummed mailing tape to produce D^1U^1-tape (see Break 3). Continue folding until you have 50, or more, triangles. Throw away the first 10 triangles, and then cut off a strip containing 31 triangles.[8] This is the pattern piece you need for this model. Prepare it as described above and then place your strip so that the left-hand end appears as shown in Figure 15(a) *with the colored side visible*. Mark the first and eighth triangles *exactly* as shown (note the orientation of the various letters within their respective triangles).

Begin by placing the first triangle *over* the eighth triangle so that the corner labeled Ⓐ is over the corner labeled A, Ⓑ is over the corner labeled B, and Ⓒ is over the corner labeled C. Hold these two triangles together, in that position, and observe that you have the beginning of a double pyramid for which there will be five triangles above and five triangles below the horizontal plane of symmetry, as shown in Figure 15(b). Now you can hold the model up and let the long strip of triangles fall around this frame. If the strip is folded well, the remaining triangles will simply fall into place. When you get to the last triangle, there will be a crossing of

---

[7]Perhaps we should have told you to wear some very old, or at least washable, clothes while doing this.

[8]You can now continue folding triangles on this piece of tape to produce triangles for building other models.

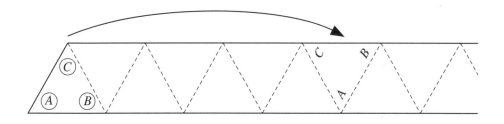

(a)

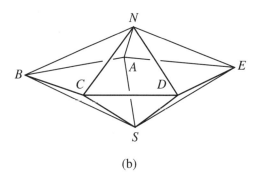

(b)

FIGURE 15

the strip that the last triangle can tuck into, and *your pentagonal dipyramid is complete!* It should look like Figure 15(b)

If you have trouble because the strip doesn't seem to fall into place, there are two most likely explanations. The first (and more likely) is that you haven't folded the crease lines firmly enough. This situation is easily remedied by refolding each crease line with more conviction. The second common difficulty occurs when the tape seems too short to reach around the model and tuck in. This problem can be remedied by trimming off a tiny amount more from each edge of the tape.

Triangular dipyramid constructed from one strip

You may wish to figure out how to make the analogous construction of a triangular dipyramid from a strip of 19 equilateral triangles. Of course, you prepare the pattern strip exactly the same way, and then knowing that the

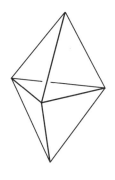

FIGURE 16

finished model should appear as shown in Figure 16 and that you should begin by forming the *top* three faces with one end of the strip should get you off to a good start.

Experiment and see if you can construct either of the above dipyramids with fewer than the number of triangles specified here. This can, in fact, be done, but the real question is: Will they be balanced, in the sense that every face is covered by the same number of triangles? This is not a difficult question to answer — and it is therefore left to the reader.

The rest of the models for which we will describe constructions are each braided from two or more straight strips of paper, and each of them is a *regular convex* polyhedron; that is, each is a model of a polyhedron known as a *Platonic Solid.*[9] We give one construction here for the tetrahedron, octahedron, and icosahedron (and suggest another, more complicated, but arguably more symmetric, construction in Chapter 8). We give two constructions here for the cube and the dodecahedron. Here are the details.

Tetrahedron constructed from 2 strips

Prepare two strips of 5 triangles each, as shown in Figure 17(b). Then, on a flat surface, with the colored surfaces down so that they are not visible, lay one strip *over* the other strip exactly as shown in Figure 17(c). Think of triangle *ABC* as the *base* of the tetrahedron being formed; for the moment, triangle *ABC* remains on the table. Then fold the bottom strip into a tetrahedron by lifting up the two triangles labeled *X* and overlapping

---

[9]One, non-technical, way of characterizing a Platonic solid is to say that it is a convex polyhedron having the property that it appears the same when it is viewed looking straight on at any vertex, or looking straight on at any edge, or looking straight on at any face. You can easily see that although the pentagonal and triagonal dipyramids are convex, they do not satisfy the rest of the conditions.

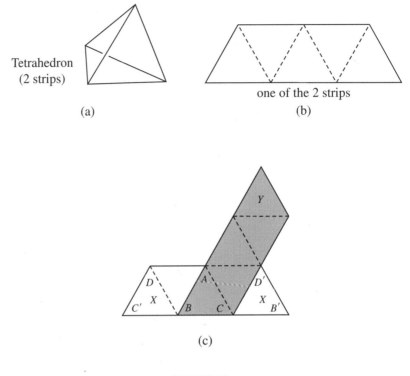

Tetrahedron
(2 strips)

(a)

one of the 2 strips

(b)

(c)

FIGURE 17

them, so that C' meets C, B' meets B, and D' meets D. Don't worry about what is happening to the top strip, as long as it stays in contact with the bottom strip where the two triangles originally overlapped. Now you will have a tetrahedron, with three triangles sticking out from one edge. Complete the model by carefully picking up the whole configuration, holding the overlapping triangles X in position, wrapping the protruding strip around two faces of the tetrahedron and tucking the Y triangle into the open slot along the edge BC. Your model should have 4 triangular faces and look like Figure 17(a), with 2 triangles from each strip visible on its surface.

A suggestion for how to build a "more symmetric" tetrahedron with three strips appears in Chapter 8.

Octahedron constructed from 4 strips.

Prepare four strips of 7 triangles each, as shown in Figure 18(b).

To construct the octahedron, begin with a pair of overlapping strips held together with a paper clip with the colored side visible, as indicated in Figure 19(a). Fold these two strips into a double pyramid by placing

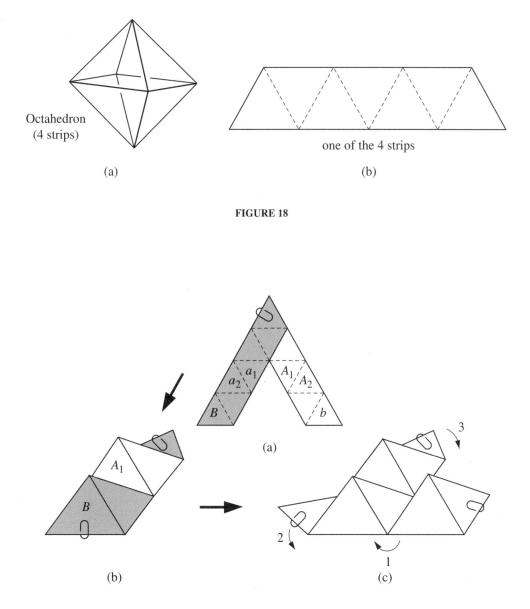

Octahedron
(4 strips)

one of the 4 strips

(a)

(b)

FIGURE 18

(a)

(b)

(c)

FIGURE 19

triangle a_1 under triangle A_1, and triangle b under triangle B. The overlapping triangles b and B are secured with another paper clip, so that the configuration looks like Figure 19(b). Repeat this process with the second pair of strips, and place the second pair of braided strips over the first pair, as shown in Figure 19(c). When doing this, make certain the flaps with the paper clips are oriented exactly as shown. Complete the octahedron by following the steps indicated by the arrows in Figure 19(c). You will note that after step 1 you have formed an octahedron; performing step 2 simply places the flap with the paper clip on it against a face of the octahedron; in step 3 you should tuck the flap *inside the model*.

When you become adept at this process you will be able to slip the paper clips off as you perform these last three steps — but they won't show, so this is only an aesthetic consideration. Your finished model should have 8 triangular faces and look like Figure 18(a), with 2 triangles from each strip visible on its surface.

A suggestion for how to build a "more symmetric" octahedron, by putting slits in the triangles of each strip, is given in Chapter 8.

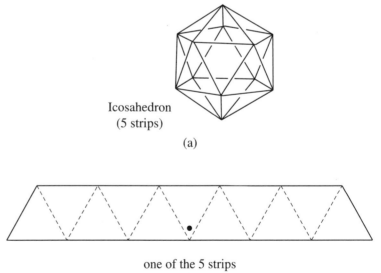

Icosahedron
(5 strips)

(a)

one of the 5 strips

(b)

FIGURE 20

Icosahedron constructed from 5 strips[10]

Prepare five strips of 11 triangles each, as shown in Figure 20(b). Lay the pieces down so that the colors are not visible and mark a heavy dot on the center triangle of each strip, as shown in Figure 20(b).

Then (without moving the pattern pieces) label one of the 5 strips with a "1" on each of its 11 triangles (make sure you are writing on the surface that will be on the *inside* of the finished model). Then label the next strip

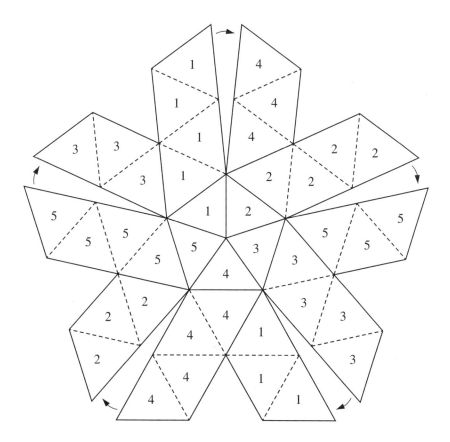

FIGURE 21

[10]Of all the models described in this section, this is, by far, the most difficult to build. So tackle this one only when you have plenty of time and patience available.

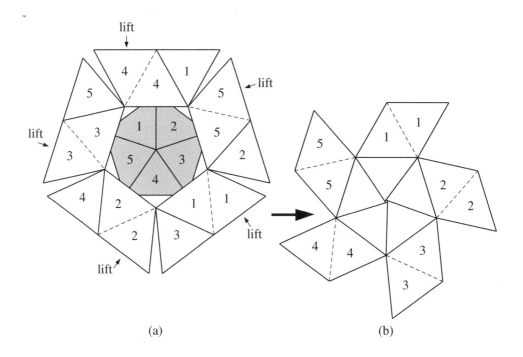

(a)   (b)

FIGURE 22

with a "2" on each of its triangles, the next with a "3", the next with a "4", and, finally, the last with a "5".

Now lay the five strips out so that they overlap each other *precisely* as shown in Figure 21, making sure that the center five triangles form a shallow cup that points *away* from you. If you do this correctly, all five dots will be hidden. You may wish to use some transparent tape to hold the strips in this position. If you do need the tape, it works best to put a small strip along the middle of each of the five lines coming from the center of the configuration (this tape won't show when the model is finished).

Now study the situation carefully before making your next move. You must bring the ten ends up so that the part of the strip at the tail of the arrow goes under the part of the strip at the head of the arrow (this means "under" as you look down on the diagram; it is really on the outside of the model you are creating, because we are looking at the inside of the model). Half the strips wrap in a clockwise direction, and the other end of each of those strips wraps in a counterclockwise direction. What finally happens is that each strip overlaps itself at the top of the model. But, in

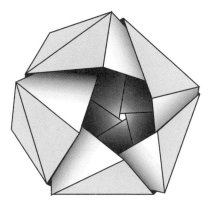

FIGURE 23

the intermediate stage, the model will look like Figure 22(a). At this point it may be useful to slide a rubber band, from the bottom, up around the emerging polyhedron to just below where the flaps are sticking out from the open pentagon (be careful not to use a rubber band that fits too tightly). Then lift the flaps as indicated by the arrows and bring them toward the center so that they tuck in, as shown in Figure 22(b).

Now simply lift flap 1 and smooth it into position. Do the same with flaps 2, 3, and 4. Complete the model by tucking flap 5 into the obvious slot. The vertex of the icosahedron nearest you will look like Figure 23.

Your finished model should have 20 triangular faces and look like Figure 20(a), with 4 triangles from each strip visible on its surface.

Congratulations! You have completed the most difficult model in this section.

A suggestion for how to build a "more symmetric" icosahedron with 6 strips appears in Chapter 8.

Cube constructed from 3 strips

Prepare three strips of 5 squares each, as shown in Figure 24(b). Figure 24(c) shows one possible set of exact fold lines that produces the desired 5 squares. Be sure only to fold on the *short* lines of the tape after you cut out the pattern piece.

The construction may be accomplished by first taking one strip and clipping the end squares together with a paper clip. Then take a second strip and wrap it around the outside of the "cube" so that one square covers the clipped squares from the first strip and the end squares cover one of

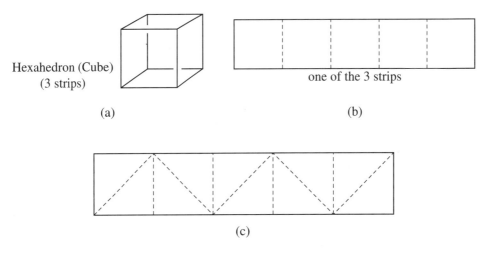

Hexahedron (Cube)
(3 strips)

one of the 3 strips

(a)

(b)

(c)

FIGURE 24

the square holes. Secure this second strip with a paper clip. Make certain
that the overlapping squares of the second strip do not cover any squares
from the first strip and that the first paper clip is covered. The pieces should
now appear as shown in Figure 25(a). Now slide the third strip under the
top square so that two squares of the third strip stick out on both the right
and left sides of the cube. Tuck the end squares of this third strip inside
the model through the slits along the bottom of the cube, as indicated in
Figure 25(b). When the completed cube of Figure 25(b) is turned upside
down, it may be opened by pulling up on the strip that covers the top face
(this square will be attached inside the model by a paper clip, so you may
have to pull firmly) and folding back the flaps that were the last to be tucked

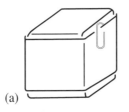

(a)

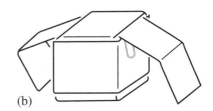

(b)

FIGURE 25

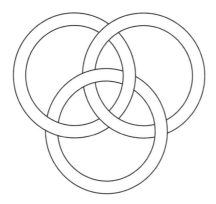

FIGURE 26

inside the model. You can remove the paper clips and put the model back together. Friction should hold the strips in place.

The cube, constructed as described above, will have 6 square faces, with opposite faces of the same color. Furthermore, if you remove any one of the strips, the other two will fall free — this is just the situation with the well-known Borromean rings, shown in Figure 26. It is also possible to construct a cube so that there are three pairs of adjacent faces having the same color. We leave this as an exercise for the interested reader.

Cube constructed from 4 strips

Prepare four strips of 7 right isosceles triangles as shown in Figure 27(a). Consult Figure 24(c) to see how to make exact folds on the tape in order to produce the pattern pieces. This time, however, you will need a longer piece of the tape, and you will only fold on the *long* lines after you cut out each pattern piece.

Begin the construction by laying the 4 strips on a table with the colored side *down*, exactly as shown in Figure 27(b). The first time you do this it may be helpful to secure the center (where the 4 strips cross each other) with some transparent tape. It is sometimes useful to put a can of unopened soda pop on this square to hold the arrangement in place and allow you to work on the vertical faces. (Remove the can before you complete the top face!) Now think of the center square in Figure 27(b) as the base of the cube you are constructing and note that the strip near the tail of each arrow should go *under* the strip at the head of the arrow (thus the strip near the tail will be on the outside of the model when it crosses the vertical edge of the

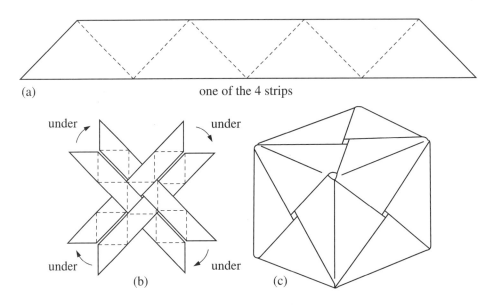

(a)   one of the 4 strips

(b)   (c)

FIGURE 27

cube). The procedure for completing the cube is now almost self-evident, especially if you remember that every strip must go alternately over and under the other strips on the model. It may help to secure the centers of the vertical faces with transparent tape as you complete them, but as you become experienced at this construction, you will soon abandon such aids. The final triangles will tuck in to produce the cube of Figure 27(c).

This cube, which we naturally call the ***diagonal cube***, has some remarkable combinatorial properties. For example, how many ways can you arrange 4 colors in a circle? Look at the faces of this cube. How many ways can you take 3 of 4 colors and arrange them in a circle? Look at the vertices of this cube. How many ways can you take 4 colors 2 at a time? Look at the edges of the cube that are opposite each other, with respect to the center of the cube.

There are other remarkable facts connected with the diagonal cube that are discussed in Chapter 8.

Dodecahedron constructed from 6 strips

Prepare 6 strips from the D^2U^2- tape so that you have 3 pairs of strips like the pair shown in Figure 28(b).

Notice that in preparing the pattern pieces for the dodecahedron shown in Figure 28(a) we only need to use the short lines on the D^2U^2-tape, but

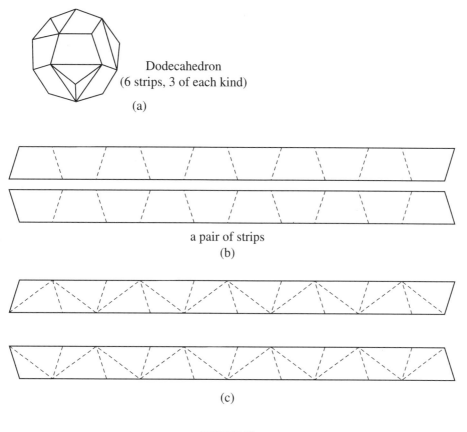

Dodecahedron
(6 strips, 3 of each kind)

(a)

a pair of strips

(b)

(c)

FIGURE 28

the gummed tape will look like that shown in Figure 28(c). Thus, after you glue the gummed tape to the colored paper you should only refold on the short lines after you cut out the pieces.

To construct this model take one pair of the strips and cross them with the colors visible as shown in Figure 29.

Secure the overlapping edge with a bobby pin (a paper clip will *NOT* work) or stick some transparent tape along each of the edges that are within the two center pentagons. Then make a bracelet out of each of the strips in such a way that

1. four sections of each strip overlap, and
2. the strip that is *under* on one side of the bracelet is *over* on the other side of the bracelet. (This will be true for both strips.)

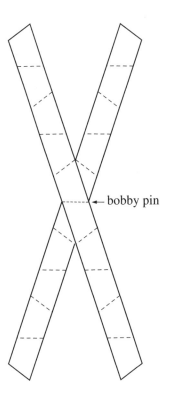

FIGURE 29

Use another bobby pin to hold all four thicknesses of tape together on the edge that is opposite the one already secured by a bobby pin.

Repeat the above steps with another pair of strips. You now have two identical bracelet-like arrangements. Slip one inside the other one as illustrated in Figure 30, so that it looks like a dodecahedron with triangular holes in four faces.

Take the last two strips and cross them precisely as you did earlier (reversing the crossing would destroy some of the symmetry on the finished model); then secure them with a bobby pin. Carefully put two of the loose ends (either the top two or the bottom two) through the top hole and pull them out the other side so that the bobby pin lands on *CD*. Then put the other two ends through the bottom hole and pull them out the other side (see Figure 31(a)). Now you can tuck in the loose flaps, but make certain to reverse the order on the strips — that is, whichever one was *under* at *CD* should be on *top* when you do the final tucking (and, of course, the top

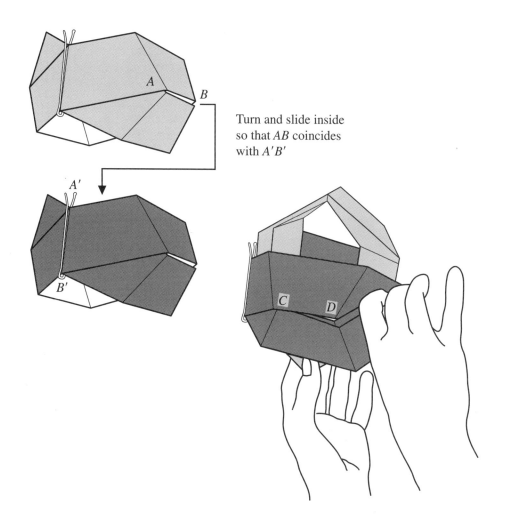

Turn and slide inside
so that *AB* coincides
with *A′B′*

FIGURE 30

strip at *CD* will be the bottom strip when you do the tucking). Your model
should look like Figure 28(a), with each face having two colors on it.

After you have mastered this construction you may wish to try to construct the model with tricolored faces, shown in Figure 31(b). This construction and the one just described are both very similar to the construction
for the cube with 3 strips. The difference is that in the case of the dodecahedron, the three "bracelets" that are braided together are each composed of
two strips. This illustrates, rather vividly, exactly how to inscribe the cube

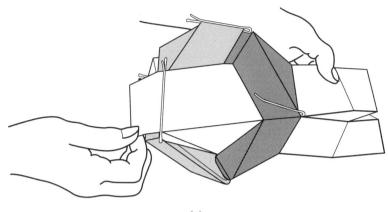

(a)

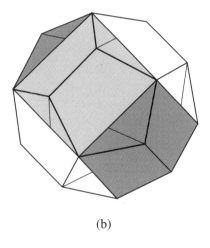

(b)

FIGURE 31

symmetrically inside the dodecahedron. To put it another way, it shows
how the dodecahedron may be constructed form the cube by placing a
"hip roof" on each of the 6 faces of the cube. You should be able to see
exactly what the hip roof looks like by examining the dodecahedron with
two colors on each face.

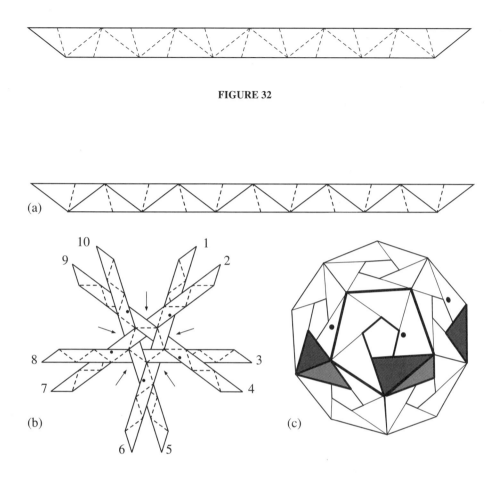

FIGURE 32

FIGURE 33

Golden Dodecahedron constructed from 6 strips

Prepare 6 strips from the D^2U^2- tape so that each strip has 22 triangles, exactly as shown in Figure 32.

When you have cut out the pattern pieces remember that, this time, you should refold each piece only on the *long* fold lines, so that the mountain fold is on the colored side of the strip. Leave the short lines *uncreased*, so that each of your 6 strips looks like Figure 33(a).

To complete the construction, begin by taking five of the strips and arranging them, with the colors visible, as shown in Figure 33(b). Secure this arrangement with paper clips at the points marked with arrows. View the center of the configuration as the North Pole. Lift this arrangement and slide the even-numbered ends clockwise over the odd-numbered ends to form the five edges coming south from the arctic pentagon. Secure the strips with paper clips at the points indicated by dots (notice where these dots are located on the finished model in Figure 33(c)). Now weave in the sixth (equatorial) strip, shown shaded in Figure 33(c), and continue braiding and clipping, where necessary, until the ends of the first five strips are tucked in securely around the South Pole. During this last phase of the construction, *keep calm and take your time!* Just make certain that every strip goes alternately over and under every other strip all the way around the model. When the model is complete, all the paper clips may be removed, and the model will remain stable.

We think this model is remarkably elegant, and it has lovely symmetry. We have named it the Golden Dodecahedron because the ratio of the length of the long line to the length of the short line on the D^2U^2-tape is, in fact, the golden ratio. See Chapter 8 for more discussion about this model and its symmetry.

★ **4.5 THE GENERAL QUASI-ORDER THEOREM**

Now, back to mathematics! In Section 2 we obtained a universal algorithm for folding a $\{\frac{b}{a}\}$-gon, where a, b are coprime odd integers with $a < \frac{b}{2}$. But, from the number-theoretic point of view, it turns out that we have much more. For, from our definition of the *symbol* (10) associated with a given set of folding instructions for our tape, we were able (in [2]) to state and prove

Theorem 2   (*The Quasi-Order Theorem in base 2*)   *If a, b are coprime integers with $a < \frac{b}{2}$, and if*

$$b \begin{vmatrix} a_1 & a_2 & \cdots & a_r \\ k_1 & k_2 & \cdots & k_r \end{vmatrix}, \qquad a_1 = a, \quad a_i \neq a_j \quad \text{if} \quad i \neq j, \quad (17)$$

*with $b - a_i = 2^{k_i} a_{i+1}$, $i = 1, 2, \cdots, r$, $(a_{r+1} = a_1)$ is a **reduced** and **contracted** symbol, and if $k = \sum_{i=1}^{r} k_i$, then k is the quasi-order of 2 mod b and, indeed,*

$$2^k \equiv (-1)^r \bmod b \qquad (18)$$

Here the **quasi-order** of t mod b, where t, b are coprime, is the smallest positive integer k such that[11]

$$t^k \equiv \pm 1 \bmod b \tag{19}$$

Before we pursue the generalization of this theorem to a general base t we will give you an example and let you experiment with some particularly interesting numbers in a break.

Example 1 **(continued)** Notice that we could find $k = 5$ from any part of the complete symbol for 31 (see item (9)). Also note that r is an even number in each of the three parts of the symbol. Thus the Quasi-Order Theorem tells us that $k = 5$ is the smallest positive integer such that $2^k \equiv \pm 1 \bmod 31$ and, moreover, that

$$2^k \equiv 1 \bmod 31$$

It is certainly easy to confirm, in this case, that this is true.

● ● ● **BREAK 7**

(1) Use your result from Break 5 (2) to find out what the Quasi-Order Theorem tells you when $b = 91$.
(2) A number of the form $2^p - 1$, where p is prime, is called a **Mersenne number** because Abbé Mersenne studied these numbers in his search for primes. Construct the symbol (17) for $b = 23$ (with $a = 1$) and see what this tells you about one of the Mersenne numbers. How do you think Mersenne would have liked this result?
(3) Recall that Fermat expected that all numbers of the form $F_n = 2^{2^n} + 1$ would be prime. Construct the symbol (17) for $b = 641$ (with $a = 1$) and see what this tells you about F_5. Notice that this result is achieved without ever calculating with numbers greater than 641.

You may have inferred from the above examples that the symbol (17) gives us, in each case, only one factor of $2^k \pm 1$. However, we described in Section 4.6 of [2] how the symbol could be used to obtain the *complementary* factor. (It is just A_1 as in item (38), p. 135 of [2], since $a = 1$.)

Now we return to our main question: *how do we generalize the Quasi-Order Theorem?* It is interesting, and not altogether surprising, that our main difficulty in generalizing this theorem to a general base t lies not in *proving* the generalization but in *stating* it. For generalization is an art, not

---

[11]It is not very difficult to prove that such a number k always exists.

an algorithmic procedure, so judicious choices must be made in formulating the generalization. Let us therefore first recall how the symbol (17) was constructed.

We start with an odd positive integer b and an odd positive integer $a < \frac{b}{2}$; at this stage we do *not* insist that a is prime to b. We then execute the **Ψ-algorithm** as follows. We choose k to be maximal such that $2^k | b - a$, so that $k \geq 1$, and set $\Psi(a) = a'$, where $b = a + 2^k a'$. We then show that Ψ is a permutation of the set S_b of odd positive integers $a < \frac{b}{2}$ which preserves $\gcd(a, b)$; that is, $\gcd(a', b) = \gcd(a, b)$ if $\Psi(a) = a'$. As a tool in proving that Ψ is a permutation, we introduce the **reverse algorithm Φ**. Thus, given $c \in S_b$, we proceed as follows: we choose ℓ minimal such that $2^\ell c > \frac{b}{2}$, so that $\ell \geq 1$, and we set $c' = b - 2^\ell c$. Then $\Phi(c) = c'$. We may then prove that Φ is the *inverse* of Ψ; that is, $\Phi(a') = a$ if and only if $\Psi(a) = a'$. Do you see why c' belongs to S_b?

Now, given b, a we construct (17) by setting $\Psi(a_i) = a_{i+1}, i = 1, \ldots, r$, *where $a_{r+1} = a_1 = a$*; more precisely,

$$b = a_i + 2^{k_i} a_{i+1}, \quad i = 1, \cdots, r \tag{20}$$

Note that, since Ψ is a permutation of S_b, we must eventually find r such that $a_{r+1} = a_1$; at that point we stop.

It is thus the Ψ- and Φ-algorithms that must be generalized. We describe how this is done so that you may see that we are indeed generalizing to a general base $t \geq 2$ from the special case $t = 2$, although the choice of generalization is not always obvious. It turns out to be easier to generalize the Φ-algorithm than the Ψ-algorithm, so we tackle that first.

We start with a positive integer b prime to t and we choose a positive integer $c < \frac{b}{2}$ such that $t \nmid c$. (Notice that the original restriction that c be odd generalizes differently from the condition that b be odd; and that the "generalization" of the condition $c < \frac{b}{2}$ is precisely $c < \frac{b}{2}$.) We write $c \in S_b$, where now S_b stands for the set of integers $< \frac{b}{2}$ that are not divisible by t, and define ℓ to be minimal for the property

$$t^\ell c > \frac{b}{2} \tag{21}$$

Notice that $\ell \geq 1$. We now define qb to be the *integer multiple of b nearest to $t^\ell c$*. Notice that

(i)　when $t = 2$ we always have $q = 1$ and $2^\ell c < b$;

(ii)　there is a *unique* such multiple. For $t^\ell c$ cannot be a multiple of $\frac{b}{2}$, because this would imply that $t^\ell c = \frac{sb}{2}$, for some s, so $sb = 2t^\ell c$, $b | 2t^\ell c$. But b, t are coprime, so $b | 2c$, contradicting $c < \frac{b}{2}$.

We now introduce a quantity ϵ such that $\epsilon = 0$ or $\epsilon = 1$. If $t^\ell c > qb$, then $\epsilon = 0$; and if $t^\ell c < qb$, then $\epsilon = 1$. In any case, we define $c' > 0$ by

$$t^\ell c - qb = (-1)^\epsilon c', \qquad (22)$$

and set

$$\Phi(c) = c'. \qquad (23)$$

Notice that, as stated earlier, with $t = 2$, we always have $q = 1$, $2^\ell c < b$, so $\epsilon = 1$. We now claim

Theorem 3   *The algorithm Φ, defined above, is a permutation of S_b such that* $\gcd(c, b) = \gcd(c', b)$, *where* $\Phi(c) = c'$.

Proof   First we must show that $c' \in S_b$. It is obvious from the definition of Φ that $c' < \frac{b}{2}$. Now $t^{\ell-1}c < \frac{b}{2}$, so $t^\ell c < \frac{tb}{2}$. Thus (see Figure 34)

$$\text{if } t \text{ is odd,} \quad 2q + 1 \leq t; \qquad \text{if } t \text{ is even,} \quad 2q \leq t. \qquad (24)$$

In either case, (24) shows that $t \nmid q$. But if $t \mid c'$ then, from (22), $t \mid qb$, so that t, b being coprime, $t \mid q$. Hence, as required, $t \nmid c'$, so $c' \in S_b$, and Φ maps S_b to itself.

• • •   BREAK 8

Draw the figure corresponding to Figure 34 for t even.

It remains to show that Φ is a permutation of S_b. Thus we seek another function $\Psi : S_b \to S_b$, inverse to Φ, that is, such that

$$\Phi\Psi = \text{Id}, \quad \Psi\Phi = \text{Id}, \qquad (25)$$

where Id represents the appropriate identity function.

However, it is of great practical importance to point out that each of the relations in (25) implies the other (after all, mathematicians strongly dislike doing unnecessary work!). For if, say, $\Phi\Psi = \text{Id}$, then

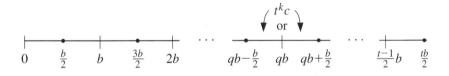

FIGURE 34   The Φ-algorithm with t odd.

Φ maps S_b *onto* itself. But S_b is a *finite* set, so Φ must be a one-one correspondence, with inverse Ψ.

We now proceed to define Ψ. Here we must distinguish between the two cases (i) t odd, (ii) t even.

(i) **t odd.** Let $a \in S_b$. We claim that, among the $(t - 1)$ positive integers

$$\left\{ qb - a, \quad qb + a, \quad 1 \le q \le \frac{t - 1}{2} \right\}$$

there is *exactly one* divisible by t. For, trivially, the t integers $-\frac{t-1}{2} \le r \le \frac{t-1}{2}$ run through all residue classes mod t. Therefore, since b is prime to t, so do the t integers rb with r in the given range; and so too, therefore, do the t integers $rb + a$ with r in the given range. However, $r = 0$ does not yield the residue class 0, since $t \nmid a$. Thus, to obtain $rb \mid a = 0 \bmod t$, we must take r strictly positive or negative. If $r > 0$, set $r = q$. Then, for one value of q in $1 \le q \le \frac{t-1}{2}$, we have $qb + a \equiv 0 \bmod t$. If $r < 0$, set $r = -q$. Then, for one value of q in $1 \le q \le \frac{t-1}{2}$, we have $qb - a \equiv 0 \bmod t$. Moreover, if there is a value of q yielding $qb + a \equiv 0 \bmod t$, there is no value of q yielding $qb - a \equiv 0 \bmod t$, and conversely. This establishes our claim.

(ii) **t even.** Let $a \in S_b$. We claim that, among the $(t - 1)$ positive integers

$$\left\{ qb + a, \quad 1 \le q \le \frac{t}{2} - 1; \quad qb - a, \quad 1 \le q \le \frac{t}{2} \right\}$$

there is *exactly one* divisible by t. Now we start with the range $-\frac{t}{2} \le r \le \frac{t}{2} - 1$ and proceed just as in the case when t is odd. We confidently leave the details to the reader.

We now complete the definition of Ψ, but will be content to describe Ψ explicitly only when t is *odd*. We expect the reader to supply the modification needed when t is even.[12] We choose q as explained above, and define a' by the rule

$$qb \pm a = t^k a', \qquad k \quad \text{maximal} \quad (\text{so that} \quad k \ge 1) \qquad (26)$$

---

[12]This is an important exercise, since we would wish to be sure that we are generalizing the case $t = 2$.

We claim that $a' \in S_b$. Now obviously, $t \nmid a'$, by the maximality of k. Also

$$ta' \leq t^k a' = qb \pm a \leq \frac{t-1}{2}b \pm a < \frac{tb}{2} \tag{27}$$

so that $a' < \frac{b}{2}$. We set $\Psi(a) = a'$.

We now prove that $\Phi\Psi = \mathrm{Id}$; recall that this will establish that Φ, Ψ are mutually inverse permutations of S_b. Assume t odd, and let $\Psi(a) = a'$. We want to show that $\Phi(a') = a$. Now, from (27), we easily see that $t^{k-1}a' < \frac{b}{2}$, and from (26) we easily see that

$$t^k a' - qb = \pm a$$

This establishes that $\Phi(a') = a$, so $\Phi\Psi = \mathrm{Id}$ as claimed. The proof of Theorem 3 will be complete (but, of course, we have proved much more) if we show that $\gcd(c, b) = \gcd(c', b)$, or, equivalently, that $\gcd(a, b) = \gcd(a', b)$, where $\Psi(a) = a'$. We choose to prove the former.

Now we have $t^k c - qb = (-1)^\epsilon c'$. Thus, if $d|c, d|b$, then $d|c', d|b$. Conversely, if $d|c', d|b$, then $d|t^k c, d|b$. But, since t, b are coprime, it follows that if $d|b$, then d is prime to t, so $d|c$. We have thus established that $\gcd(c, b) = \gcd(c', b)$. $\qquad\square$

In describing the Φ-algorithm, we introduced the quantity ϵ, which takes the value 0 or 1 according to whether $t^k c - qb$ is positive or negative (see (22)). For consistency we must introduce it again in (26), where we must choose between $qb + a$ or $qb - a$, when seeking the allowed number divisible by t. Thus we refine (26) to

$$qb + (-1)^\epsilon a = t^k a' \tag{28}$$

We are now in a position to define a ***t-symbol***, which we will still just call a symbol if there is no doubt what base t is being used. However, we will add an extra row to the symbol, by comparison with the case $t = 2$, to incorporate the quantity ϵ (recall that if $t = 2$, we always have $\epsilon = 1$). On the other hand, we will not need to include the quantity q from (28) in the symbol if our purpose is just to state and prove the general Quasi-Order Theorem.

We suppose that b, a are coprime, with b prime to t, $t \nmid a$, and $a_1 = a < \frac{b}{2}$. We proceed to obtain a contracted, reduced symbol

$$
b \begin{vmatrix} a_1 & a_2 & \cdots & a_r \\ k_1 & k_2 & \cdots & k_r \\ \epsilon_1 & \epsilon_2 & \cdots & \epsilon_r \end{vmatrix}_t \tag{29}
$$

where

$$
q_i b + (-1)^{\epsilon_i} a_i = t^{k_i} a_{i+1}, \quad i = 1, 2, \cdots, r \quad (a_{r+1} = a_r) \tag{30}
$$

and there are no repeats among the a_i, $i = 1, 2, \cdots, r$; recall that we know that each a_i is prime to b, $t \nmid a_i$, and $a_i < \frac{b}{2}$. Our symbol (29) is, of course, based on the Ψ-algorithm.

Notice that, if we use the Φ-algorithm instead of the Ψ-algorithm, we get the *reverse symbol*, incorporating the same r facts (30); thus, we may replace (29) by

$$
b \begin{pmatrix} a_1 & a_r & a_{r-1} & \cdots & a_2 \\ k_r & k_{r-1} & k_{r-2} & \cdots & k_1 \\ \epsilon_r & \epsilon_{r-1} & \epsilon_{r-2} & \cdots & \epsilon_1 \end{pmatrix}_t \tag{31}
$$

For neatness, we rewrite (31) as

$$
b \begin{pmatrix} c_1 & c_2 & \cdots & c_r \\ \ell_1 & \ell_2 & \cdots & \ell_r \\ \eta_1 & \eta_2 & \cdots & \eta_r \end{pmatrix}_t \tag{32}
$$

so that $c_1 = a_1$, $c_j = a_{r+2-j}$, $2 \leq j \leq r$, $\ell_j = k_{r+1-j}$, $\eta_j = \epsilon_{r+1-j}$, and (30) is to be rewritten as

$$
q'_j b + (-1)^{\eta_j} c_{j+1} = t^{\ell_j} c_j \tag{33}
$$

(where, in fact, $q'_j = q_{r+1-j}$).

We set $L = \sum \ell_j$, $E = \sum \eta_j$ from (32); of course, it is equally true that $L = \sum k_i$, $E = \sum \epsilon_i$ from the symbol (29).

Example 4   Form the symbol (29) with $b = 19$, $a_1 = 6$, $t = 4$.

Solution   From the following calculations we obtain the symbol below:

$$2 \cdot 19 - 6 = 4^2 \cdot 2$$
$$2 \cdot 19 - 2 = 4^1 \cdot 9$$
$$19 + 9 = 4^1 \cdot 7$$
$$19 - 7 = 4^1 \cdot 3$$
$$19 - 3 = 4^2 \cdot 1$$
$$19 + 1 = 4^1 \cdot 5$$
$$19 + 5 = 4^1 \cdot 6$$

$$
\left. 19 \,\middle|\,
\begin{array}{ccccccc}
6 & 2 & 9 & 7 & 3 & 1 & 5 \\
2 & 1 & 1 & 1 & 2 & 1 & 1 \\
1 & 1 & 0 & 1 & 1 & 0 & 0
\end{array}
\right|_4
$$

Example 5   Form the symbol (29) with $b = 19$, $a_1 = 6$, $t = 5$.

Solution   From the following calculations we obtain the symbol below:

$$19 + 6 = 5^2 \cdot 1$$
$$19 + 1 = 5^1 \cdot 4$$
$$19 - 4 = 5^1 \cdot 3$$
$$2 \cdot 19 - 3 = 5^1 \cdot 7$$
$$2 \cdot 19 + 7 = 5^1 \cdot 9$$
$$19 - 9 = 5^1 \cdot 2$$
$$2 \cdot 19 + 2 = 5^1 \cdot 8$$
$$2 \cdot 19 - 8 = 5^1 \cdot 6$$

$$
\left. 19 \,\middle|\,
\begin{array}{cccccccc}
6 & 1 & 4 & 3 & 7 & 9 & 2 & 8 \\
2 & 1 & 1 & 1 & 1 & 1 & 1 & 1 \\
0 & 0 & 1 & 1 & 0 & 1 & 0 & 1
\end{array}
\right|_5
$$

Now, let's take a break so you can get some practice!

• • • **BREAK 9**

(1) Form the symbol with $b = 13$, $a_1 = 6$, $t = 4$, using the Ψ-algorithm. (Hint: In calculating $\Psi(a)$ we must find out which of $b - a$, $b + a$, $2b - a$ is divisible by 4.)

(2) Form the symbol with $b = 17$, $a_1 = 3$, $t = 5$, using the Ψ-algorithm. (Hint: In calculating $\Psi(a)$ we must find out which of $b - a$, $b + a$, $2b - a$, $2b + a$ is divisible by 5.)

We are now ready to prove the general Quasi-Order Theorem. We suppose given a Φ-symbol (32); of course, we might also suppose given the associated Ψ-symbol (29), so that the quantities L, E may be read off from either symbol as the sum of the second row and the sum of the third row, respectively. Our theorem is then the following.

Theorem 6 (***The General Quasi-Order Theorem***)   *Given a (contracted, reduced) Φ-symbol (32), then the quasi-order of t mod b is $L = \sum \ell_j$. Indeed,*

$$t^L \equiv (-1)^E \bmod b, \quad where \quad E = \sum \eta_j.$$

Proof   We consider the sequence of $L + 1$ integers

$$c_1, tc_1, \ldots, t^{\ell_i - 1}c_1, c_2, tc_2, \ldots, t^{\ell_2 - 1}c_2, c_3, \ldots, c_r, tc_r, \ldots, t^{\ell_r - 1}c_r, c_1 \tag{34}$$

We say that a ***switch*** takes place when we pass from $t^{\ell_i - 1}c_i$ to c_{i+1}, $i = 1, 2, \cdots, r$ ($c_{r+1} = c_1$). We also write (34) as

$$u_1, u_2, \ldots, u_{L+1} \tag{35}$$

We note that, following rule (21) for defining the Φ-algorithm,

$$u_j < \frac{b}{2}, \quad 1 \leq j \leq L + 1 \tag{36}$$

We also note from (33) that

$$\begin{cases} u_{j+1} = tu_j & \text{if no switch occurs,} \\ u_{j+1} \equiv (-1)^{\eta_i} tu_j \bmod b & \text{if a switch occurs to } c_{i+1}. \end{cases} \tag{37}$$

It follows that

$$u_{L+1} \equiv (-1)^E t^L u_1 \bmod b$$

or

$$c_1 \equiv (-1)^E t^L c_1 \bmod b$$

But c_1, b are coprime, so $t^L \equiv (-1)^E \bmod b$, as claimed.

It remains to show that L is indeed the quasi-order of t mod b. Suppose not; then there exists an M, where $1 \leq M < L$, with $t^M \equiv \pm 1$ mod b. Thus, from (37),

$$u_{M+1} \equiv \pm u_1 \text{ mod } b \tag{38}$$

We first show that $u_{M+1} \not\equiv u_1$ mod b. For, by (36), if $u_{M+1} \equiv u_1$ mod b, then $u_{M+1} = u_1$. If u_{M+1} arose at a switch, then, remembering that $M + 1 < L + 1$, we see that $u_{M+1} = u_1$ contradicts the fact that the symbol (32) is contracted. If u_{M+1} did not arise at a switch, then $t | u_{M+1}$, $t \nmid u_1$, again contradicting $u_{M+1} = u_1$. Hence $u_{M+1} \not\equiv u_1$ mod b.

Finally, we show that $u_{M+1} \not\equiv -u_1$ mod b. For if

$$u_{M+1} \equiv -u_1 \text{ mod } b,$$

then $b | (u_{M+1} + u_1)$. But, by (36), $u_{M+1} + u_1 < b$, so this is impossible. Thus (38) is false, and L is, as claimed, the quasi-order of t mod b. $\square$

Corollary 7 (*Alternative form of the General Quasi-Order Theorem*) *Given a (contracted, reduced) Ψ-symbol (29), then the quasi-order of t mod b is $K = \sum k_i$. Indeed,*

$$t^K \equiv (-1)^E \text{ mod } b$$

where $E = \sum \epsilon_i$.

Remark Recall our claim that the *proof* of Theorem 3 is scarcely more difficult in the general case than the special case $t = 2$; the only (slight) complication arises from the fact that η_j in (32), (33) may be 0 or 1 in the general case, whereas we always have $\eta_j = 1$ if $t = 2$. On the other hand, we do not expect you to doubt that preparing the ground for Theorem 3, that is, defining the Φ- and Ψ-algorithms, was *much* more difficult in the general case!

Example 4 (**revisited**) From the symbol we created in Example 4 we see that, referring to Corollary 7, $K = 9$, $E = 4$, so that the quasi-order of 4 mod 19 is 9, and indeed

$$4^9 \equiv (-1)^4 \text{ mod } 19 = 1 \text{ mod } 19$$

Example 5 **(revisited)** From the symbol we created in Example 5 we see that, referring to Corollary 7, $K = 9$, $E = 4$, so that the quasi-order of 5 mod 19 is 9, and indeed

$$5^9 \equiv (-1)^4 \bmod 19 = 1 \bmod 19$$

(Why do you think the values of K are the same in these two examples?)

• • • BREAK 10

(1) Refer to the Ψ-symbol you constructed in Break 9(1) and write down the Φ-symbol for $b = 13$, $c_1 = 6$, $t = 4$. From either symbol obtain the quasi-order L of 4 mod 13, and determine whether $4^L \equiv +1$ or -1 mod 13.

(2) Refer to the Ψ-symbol you constructed in Break 9(2) and write down the Φ-symbol for $b = 17$, $c_1 = 3$, $t = 5$. From either symbol obtain the quasi-order L of 5 mod 17, and determine whether $5^L \equiv +1$ or -1 mod 17.

(3) (Harder) Show that the quasi-order of t mod b, where b is an odd prime, is always a factor of $\frac{1}{2}(b - 1)$. (Hint: Use Fermat's Little Theorem (Ch. 2 of [2].))

(4) Answer the question following Example 5 using the hint given for solving question (3).

REFERENCES

1. Coxeter, H. S. M., *Regular Polytopes*, Macmillan Mathematics Paperbacks, New York (1963).

2. Hilton, Peter, Derek Holton, and Jean Pedersen, *Mathematical Reflections — In a Room With Many Mirrors*, 2nd printing, Springer-Verlag NY, 1998.

3. Hilton, Peter, and Jean Pedersen, Descartes, Euler, Poincaré, Pólya and polyhedra, *L'Enseign. Math.* **27** (1981), 327–343.

4. Hilton, Peter, and Jean Pedersen, Approximating any regular polygon by folding paper; An interplay of geometry, analysis and number theory, *Math. Mag.* **56** (1983), 141–155.

5. Hilton, Peter, and Jean Pedersen, Folding regular star polygons and number theory, *Math. Intelligencer* **7**, No. 1 (1985), 15–26.

6. Hilton, Peter, and Jean Pedersen, Certain algorithms in the practice of geometry and the theory of numbers, *Publ. Sec. Mat. Univ. Autonoma Barcelona* **29**, No. 1 (1985), 31–64.

7. Hilton, Peter, and Jean Pedersen, Geometry in Practice and Numbers in Theory, *Monographs in Undergraduate Mathematics* **16** (1987), 37 pp. (Available from Department of Mathematics, Guilford College, Greensboro, North Carolina 27410, U.S.A.)

8. Hilton, Peter, and Jean Pedersen, *Build Your Own Polyhedra*, Addison-Wesley, Menlo Park, California (1987, reprinted 1994, 1999), 175 pp.

9. Hilton, Peter, and Jean Pedersen, On the complementary factor in a new congruence algorithm, *Int. Journ. Math. and Math. Sci.* **10**, No. 1 (1987), 113–123.

10. Hilton, Peter, and Jean Pedersen, On a generalization of folding numbers, *Southeast Asian Bulletin of Mathematics* **12**, No. 1 (1988), 53–63.

11. Hilton, Peter and Jean Pedersen, On factoring $2^k \pm 1$, *The Mathematics Educator* **5**, No. 1 (1994), 29–32.

12. Hilton, Peter and Jean Pedersen, Geometry: A gateway to understanding, *The College Mathematics Journal* **24**, No. 4 (1993), 298–317.

13. Huzita, H, and B. Scimemi, The algebra of paper-folding (Origami), *Proceedings of the First International Meeting on Origami Science and Technology*, ed. by H. Huzita, Ferrara, 1989.

5

Are Four Colors Really Enough?

5.1 INTRODUCTION: A SCHOOLBOY INVENTION

There have been few problems in mathematics over the centuries that have taken the popular imagination as much as the Four Color Problem. Probably the main reason for this is that it is something that can be explained to anybody in only a minute or two. Perhaps the most surprising thing about this problem is that it was invented by a schoolboy and not by a mathematician at all. It's a very human story — we'll mention honeymoons, school challenges, and a popular magazine later. We'll also mention its 124-year history and why some people are still working on it even after it's been solved. And then there is the famous link between this problem and Lewis Carroll's poem, "The Hunting of the Snark." We'll get to that too. After reading this chapter you might like to look at the overview [1], by Appel and Haken, of their proof of the Four Color Theorem, or, for a more complete treatment, see [6], [7], or [9].

5.2 THE FOUR-COLOR PROBLEM

In 1852, a young student in England, Frederick Guthrie, was doing his geography homework. Correction: he was *supposed* to be doing his geography homework. Instead, he started doodling with a map of England which had all the counties marked on it. He started to color them in. He decided it was worth coloring the counties so that any two counties that had a common border were given a different color. So counties like Kent and Sussex, which are next to each other in the south of England, always had to be colored differently.

Frederick was able to color his map with just *four* colors. And this gave rise to the problem: is it possible to color any map in four or fewer colors so that regions with a common boundary have different colors?

This is the famous Four Color Problem. It is an unusual problem in that it was discovered by a school student and yet has caused mathematicians so much trouble. It was a problem that was to haunt mathematicians for well over 100 years, and some of them are still worried by it today, even though it has been proved that the answer is yes.

● ● ●   BREAK 1

(1) We have reproduced a map of the counties of England for you below; repeat Fred Guthrie's experiment.

(2) Invent your own set of countries by dividing a piece of paper into a finite set of regions (counties or countries). Will four colors suffice for your map? (Remember that no two adjacent regions can have the same color.)

Now it turns out that Fred had a brother called Frank, who was studying mathematics at University College, London, England. Fred told Frank, and Frank told his professor, the well-known Augustus de Morgan. No, de Morgan hadn't seen this problem before. And no, after doodling with it himself, he couldn't explain why you only needed four colors. However, he was pretty sure that you never needed a fifth color. All de Morgan's experimenting led him to the conclusion that four colors sufficed. But the fact that he couldn't justify it perplexed him somewhat. So he did what many a mathematician before and since has done, he got in touch with another mathematician. Unfortunately, he didn't have e-mail, so he wrote off to Sir William Rowan Hamilton. In that letter de Morgan gave an example of a map that actually *needed* four colors.

Hamilton, the Irish mathematician, who was responsible for some fine mathematics, including the invention of quaternions, was one of the few people who seemed not to get interested in the idea. He gave de Morgan a curt reply and got back to his research on trying to generalize complex numbers.

A map of the counties of England

● ● ● **BREAK 2**

(1) Can you find a simple map that *has* to have four colors?

(2) What are quaternions? Where are they written on a bridge and why?[1]

---

[1] See http://www.maths.tcd.ie/pub/HistMath/People/Hamilton

At this stage the Four Color Problem can really only be thought of as the Four Color Conjecture. The work on the Four Color Problem had now spread, and pretty well everyone was beginning to believe that maps only needed four colors. So the conjecture they all subscribed to was

The Four-Color Conjecture   *In coloring any map so that no two neighboring regions have the same color, at most four colors are needed.*

Despite his rebuff by Hamilton, de Morgan became the Four Color Conjecture's publicist. He even mentioned it in 1860 in the popular magazine, *The Athenaeum*. The fame of the problem had spread so far by 1886 that the headmaster of the English private school, Clifton College, issued the problem as a challenge to the school. He gave the boys a deadline by which time they should produce a solution of no more than one page of writing and one page of diagrams.

Now it is not clear whether the headmaster did this because he knew that the lawyer/mathematician Kempe had a proof, or because he knew that the proof was wrong. Kempe's "proof" is probably the most famous false proof in mathematical history. But we need a detour in order to be able to sneak up on it. Alfred Bray Kempe will reappear in Section 8.

5.3 GRAPHS

So what is a graph? In a way, it is unfortunate that in this chapter (and in Chapter 9, for that matter) a graph is not the thing that you are familiar with, something with x- and y-axes. We're interested in quite a different animal here, and, apparently, a somewhat simpler animal, too. The graphs we have to deal with have vertices, some or all of which may be joined by edges, which may be curved. But no vertex is joined to itself by an edge, and no pair of vertices is joined by more than one edge.

Have a look at Figure 1. This shows three graphs each with four vertices and five edges. Our convention is that these three graphs are the same graph in three different guises. In each case the vertices are shown as dots and the edges as lines joining the dots.

As we are going to consider all three graphs of Figure 1 to be the same, you can see that we don't worry about where we put the dots, or how plain or fancy the lines are that represent the edges. The way we'll decide whether two graphs are the same is if we can put the vertices on top of each other so that pairs of vertices that are joined by an edge always sit on top of corresponding pairs of vertices that are joined by an edge, and, moreover,

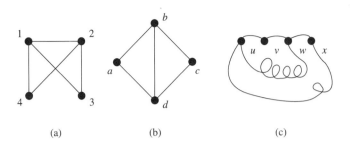

FIGURE 1   Three identical graphs

pairs of vertices that are not joined by an edge sit on top of corresponding pairs of vertices that are not joined by an edge.

Figures 1(a) and 1(b) represent the same graph because we can put the vertices 1 and b together, 2 and d together, 3 and c together, and 4 and a together. When we've done that, we can treat the edges as if they were made of elastic, and the edges 12, 13, 14, 23, and 24 sit on top of the edges bd, bc, ba, dc, and da.

In a similar way Figures 1(a) and 1(c) and Figures 1(b) and 1(c) are the same, too. We say that graphs that are the same in this way are ***isomorphic***. If A and B are isomorphic graphs, we write $A \cong B$. So we are regarding isomorphic graphs as the same, or identical.

● ● ●   **BREAK 3**

(1) Show that all the graphs of Figure 1 are isomorphic.
(2) Show that there is only one graph with one vertex, that there are two different graphs with two vertices, and four different graphs with three vertices. (Remember that you can't have an edge between a vertex and itself, and you can't have two or more edges between any two given vertices.)
(3) Use the last problem to form a conjecture about the number of graphs with n vertices. Justify the conjecture or find a counter-example.

Now, just as sets have subsets, so graphs have ***subgraphs***. If G is a graph, then H is a subgraph if (i) the vertices of H are a subset of the vertices of G and (ii) the edges of H are a subset of the edges of G. Clearly, the edges in (ii) must join vertices from (i). So the graph on vertices 1, 2, 3 with edges 12, 23 is a subgraph of the graph in Figure 1(a).

One useful concept relating to graphs is the **_degree_** of a vertex. This is simply the number of edges that go into that vertex. In Figure 1(b), vertex a has degree 2, vertex b has degree 3, vertex c has degree 2, and vertex d has degree 3. We write $\deg a = 2$, $\deg b = 3$, and so on. Notice that if H is a subgraph of G and a is a vertex of H, then the degree of a as a vertex of H may be smaller than its degree as a vertex of G.

It's also useful to describe a graph as **_connected_** if you can go between any two vertices by a sequence of vertices and edges with the edges joining consecutive vertices. Hence the graph in Figure 2(a) is connected, while the one in Figure 2(b) is not. There is no way of getting from u to v in Figure 2(b) using edges of that graph. The connected pieces of disconnected graphs are called **_components_**.

By a **_cycle_**, we mean a connected graph where each vertex has degree 2. Cycles can exist as subgraphs in which every vertex of the subgraph has degree 2. If a cycle has n vertices, then it is denoted by C_n. So in Figure 1(a), $1, 2, 3$ form a cycle, but $1, 3, 4$ don't. There are no cycles in Figure 2(a), but there are four cycles in Figure 2(b). Two of these cycles are isomorphic to C_3, and two are isomorphic to C_4. A cycle is sometimes called a **_circuit_** (see Chapter 7, Section 1.)

Now a connected graph which has no subgraphs which are cycles is called a **_tree_**. The graph in Figure 2(a) is a tree on nine vertices.

 BREAK 4

    (1) Show that no two of the following three graphs are isomorphic: C_8; the graph on eight vertices made up of two copies of C_4; the

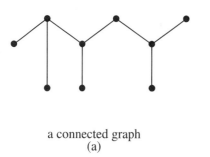

a connected graph
(a)

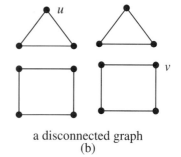

a disconnected graph
(b)

FIGURE 2

graph on eight vertices made up of one copy of C_5 and one copy of C_3.

(2) Find all trees with six vertices.

Showing that two graphs are isomorphic is a bit of a problem. In fact, the "isomorphism problem", that is, finding a general method of efficiently deciding when two graphs are isomorphic, is an unsolved problem in graph theory. We can certainly formalize the idea of putting vertices on top of one another so that edges go on top of edges and non-edges go on top of non-edges. However, it's not really worth giving a formal definition here. (See [4], [6], [9], [10], or any other graph theory book if you want to pursue this line.) It is useful to have a look at a few graphs, though, to see what makes them different. That way we'll be able to pick out graphs that are obviously *not* isomorphic. And that will be enough for us at this stage.

So have a look at the two graphs G and H in Figure 3. Are they the same, isomorphic? Probably not. They don't look very much alike. In what way do they differ, though?

● ● ● **BREAK 5**

(1) Give a number of reasons why you think G and H might not be isomorphic.

(2) The graphs G and H in Figure 3 are fairly easy to sort out. But how about the graphs K and L of Figure 4. Are they isomorphic or not? What do you think?

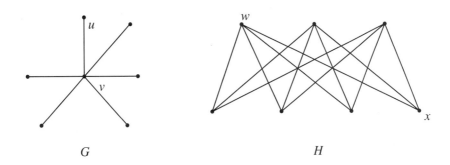

FIGURE 3

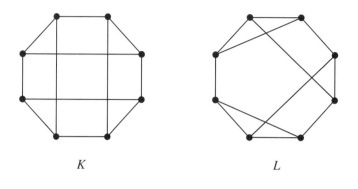

FIGURE 4

One of the things that distinguishes G from H is the number of edges. Plainly, G has only 6 edges. On the other hand, H is comparatively edge-rich with 12 edges.

In the same vein, G has some vertices of degree one. (It has six such vertices.) In fact, $\deg u = 1$ and $\deg v = 6$. It should be clear from Figure 3 and the "putting vertices on top of each other" definition of isomorphism that isomorphic graphs have corresponding vertices with the same degree.

Since G has six vertices of degree 1 and one vertex of degree 6, then any graph isomorphic to G also has to have six vertices of degree 1 and one vertex of degree 6. The graph H obviously is not one of these graphs. Hence $G \ncong H$.

So let's apply the degree test to the two graphs K and L of Figure 4. In this case, the test fails to distinguish between the graphs. Every vertex in K has degree 3, and so does every vertex in L. But is $K \cong L$? They don't look alike, or do they? Is there any way to distinguish them from each other? We'll go back to this question later, but first let's take a break.

• • • BREAK 6

In Figure 5, is $M \cong N$?

Let's go back to graphs K, L, M, and N for a moment. You've probably noticed that every vertex in these graphs has degree 3. Because of this they are usually said to be **_regular of degree_ 3**. In fact, a graph is **_regular_** if every vertex has the same degree.

You may have realized by now that regular graphs of the same degree with the same number of vertices are not necessarily isomorphic. In fact,

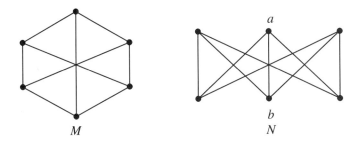

FIGURE 5

K and L are not isomorphic, although they are both regular of degree 3. One of the things that distinguishes K from L is that L has a cycle C_3, while the smallest cycle in K is C_4. So $K \not\cong L$.

On the other hand, this "cycles test" will not distinguish M from N. Actually, M and N each have one 6-cycle with all other cycles 4-cycles, so they can't be distinguished that way. In fact, there is no way to distinguish them because they are isomorphic. It's not obvious, though, just by looking at them. (But try moving vertex a of N to the bottom in Figure 5 and vertex b of N to the top.)

Among regular graphs, one important class is that of **complete** graphs, K_n. These are the graphs on n vertices in which every vertex is joined to every other vertex. We show K_2, K_3, K_4, K_5, and K_6 in Figure 6.

It's worth noting that all of them from K_3 onwards have triangles (C_3) and that all of them from K_4 onwards have 4-cycles (C_4); and so on. The other thing worth noting is that the degree of regularity of K_2 is 1, of K_3 is 2, of K_4 is 3, of K_5 is 4, and of K_6 is 5. It ought to be reasonably clear what the degree of K_n is.

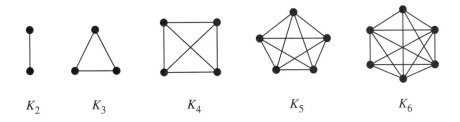

K_2 K_3 K_4 K_5 K_6

FIGURE 6

● ● ●   **BREAK 7**

That reminds us, the notion of degree came in earlier on because of its relation to edges. How many edges has K_n? And how is the number of edges related to the degree in K_n? Is there any relation between degrees and the number of edges in an arbitrary graph G?

5.4 TOURING WITH EULER

The very first problem to be tackled with the use of graphs was first discussed in the mid 1730's. That problem is the Euler Tour Problem,[2] and it all started with people walking over bridges in a place called Königsberg in what in the 1730's was Prussia (see [3]). The city is today in Russia, it's called Kaliningrad, and they've built another bridge since then.

Königsberg was a city on the banks of the river Pregel and one of its tributaries. Its approximate layout in the 1730's is shown in Figure 7.

The problem is this: Is it possible to plan a walk in which one goes over each bridge exactly once? Euler looked at the problem and modeled Figure 7 with a multigraph: this is a graph is which pairs of vertices may be joined by more than one edge. He constructed his model by placing a vertex on each separate land mass and joining two land masses by an edge for each bridge between them. Hence he produced the multigraph of Figure 8 and, in 1736, graph theory was born.

He then set to work to reason this way. If the multigraph of Figure 8 had a route that would take you across every edge (bridge) once and only once,

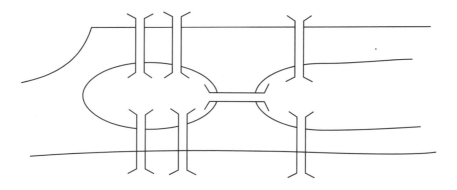

FIGURE 7

---

[2]This is popularly known, for obvious reasons, as *The Königsberg Bridge Problem*.

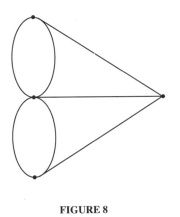

FIGURE 8

then most vertices of the graph would have to have even degree. After all, when you come to a new vertex (except possibly the last) you go into it and then out of it. So almost every time you use a vertex x (land mass), you use two edges incident with x. In your walk around Königsberg, then, using all the bridges, you have shown that every vertex except at most two has to have even degree. But none of the vertices of Figure 8 has even degree. There was therefore no way for the Königsbergians to walk over each bridge exactly once.

● ● ● **BREAK 8**

(1) What is the least number of edges that you need to add to the graph of Figure 8 so that there is a route around the graph that starts and ends at *different* vertices but uses every edge?

(2) If you want to start and finish at the *same* vertex, what is the minimum number of edges that must be added.

Being the good mathematician that he was, Euler generalized his Königsberg bridge problem to *any* multigraph. Let's say that a graph has an ***Euler tour*** if there is a way of going over every edge once and only once and returning to the starting point. Then Euler claimed the following result.

Theorem (Euler, 1736)   *A connected multigraph has an Euler tour if and only if every vertex has even degree.*

Euler proved part of this but didn't prove it all. Using the argument we used above, he showed that if a multigraph *has* an Euler tour, then every vertex *has* to have even degree. But he didn't quite manage the converse. This is not too difficult. It can be done several ways, but we outline a proof by induction on the number of edges in a multigraph.

To get the induction step to work we need to show that a multigraph G in which every vertex has even (non-zero) degree has a cycle. That argument goes like this. First, if G has a multiple edge, it has a cycle C_2, so we exclude multiple edges from here on. So let us start with any vertex v_1. Then v_1 has even degree, so v_1 is adjacent to some vertex v_2. Similarly, v_2 has even degree, so there is a vertex $v_3 \neq v_1$ that is adjacent to v_2. Then v_3 is adjacent to $v_4 \neq v_2$. If $v_4 = v_1$, we have a cycle. If not, continue. So v_4 is adjacent to $v_5 \neq v_3$. If $v_5 = v_1$ or v_2 we have a cycle. If not, continue. Since we are dealing with a finite number of vertices, we will eventually have to use one of the vertices we have already used. Hence we have a cycle.

Suppose then that we have a cycle C in G. Every vertex in C has degree 2. Remove C from G. By induction, there is an Euler tour in each of the components of the multigraph we have left. We can then patch together these Euler tours and C to give a tour of G.

● ● ●   **BREAK 9**

Suppose we want to start and end at different vertices of a multigraph and pass over every edge once and only once. Find necessary and sufficient conditions for this to be possible. Prove this.

5.5   WHY GRAPHS?

We started out worrying about ways to color the counties of England, and then we sidetracked you into thinking about graphs. Presumably, there's some relation here. What is it?

To see the advantage of graphs, we need to turn our maps "inside out." To see how to do this, look at Figure 9(a). Here we have a fairly simple map. It only has six regions (or faces), counties, or countries. They are labeled A, B, C, D, E, F.

Suppose the regions are countries. Then mark the capital city of country X with a dot labeled x and join up two capital cities if and only if their countries have a common border. This leads to the picture in Figure 9(b). Note that we always think of the "outside" region A as being a country, too.

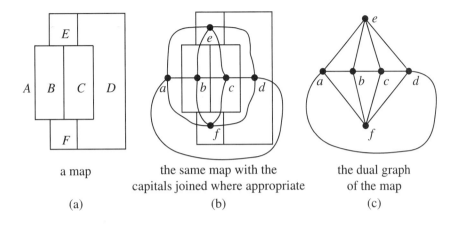

a map

the same map with the
capitals joined where appropriate

the dual graph
of the map

(a) (b) (c)

FIGURE 9

Now extract from Figure 9(b) the diagram consisting of capital cities and joins. This gives us Figure 9(c), which is a graph. Actually we call it the **dual graph** of the map.

But what has the dual graph got to do with coloring maps? Suppose our map in Figure 9(a) can be 4-colored. What does this tell us about the dual graph in Figure 9(c)? Since the vertices in some sense stand for the countries or regions, why not color the vertices of the dual graph? And we need to color them so that two adjacent vertices have different colors, because that corresponds to two countries with a common border having different colors.

Hence if we can color the *regions* of the map with four or fewer colors, we can color the *vertices* of the dual graph with four or fewer colors. Is the converse true? That is, is it true that if we can color the vertices of the dual graph in four or fewer colors, then we can color the map in the same way?

Yes, of course, this is quite straightforward. Whatever color we have on vertex a, we use over the whole of region A. Whatever color we have on vertex b, we use over the whole of region B, and so on. Now if vertex u is adjacent to vertex v they will have different colors. In that case, adjacent regions U and V will have different colors.

We've just discovered that the regions of the map can be colored in k colors so that no two neighboring regions have the same color if and only if the vertices of the dual graph can be colored in k colors so that no two adjacent vertices have the same color. Hence we've transformed the map-coloring problem into a graph-coloring problem. But of what value is that? Have we really made any progress? Is graph-coloring somehow easier to understand than map-coloring?

But maybe we're running a little ahead of ourselves. It's just possible that the map of Figure 9(a) isn't 4-colorable. Let's check it and see.

Without loss of generality let region A be colored red. Then B, D, E, and F can't be red, and since B and E are neighbors they have to have different colors. So let's assume that B is colored blue and E green.

It ought to be clear that we can save colors by letting F be colored green, too. After all, F and E are not neighbors. Now C can't be blue or green, but since it's not adjacent to A, it can be red. Then we can finish off the map by coloring D the same color as B, namely blue. It looks as if we can color this map in only three colors. And we can shift these colors over to the dual graph of Figure 9(c) by coloring a and c red, e and f green, and b and d blue.

● ● ●   **BREAK 10**

(1) All the vertices in the dual graph of Figure 9(c) have degree 4. Do all dual graphs have this property?

(2) Can the graph of Figure 9 be colored in fewer than 3 colors? How low can you go?

So how low can graph-colorings go? Can we color a graph in one color or two colors? We know that we can do some in three (see Figure 9(c)).

Just so that we know precisely what we are talking about, we'll say that a graph is **_k-colorable_** if its vertices can be colored in k colors so that no two adjacent vertices have the same color.

Going back to Figure 9(c), that graph can be colored in 6 colors. We can always give each vertex a different color so that it is 6-colorable. But we know that it can also be colored in 3 colors, so it is also 3-colorable. On the other hand, it can't be colored in 2 colors, because a, b, and e are mutually adjacent. So the graph of Figure 9(c) is _not_ 2-colorable.

Fine, but are there 1-colorable or 2-colorable graphs? What would a 1-colorable graph look like? Surely, as soon as a graph has an edge it has to have at least 2 colors. So a 1-colorable graph is just a collection of isolated vertices.

So let's try for 2-colorable graphs. Suppose the vertices are colored red and blue. If we started off from a blue vertex, we'd have to move to a red vertex. In fact, all the vertices adjacent to a blue vertex would have to be red vertices. And in the same way, all the red vertices would have to be

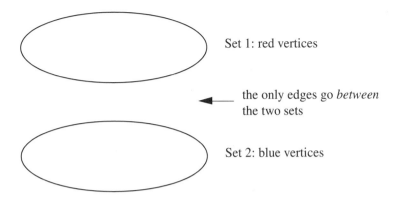

FIGURE 10   Schema for a 2-colorable graph

adjacent to blue vertices. So a 2-coloring divides the vertices into two sets, the red vertices and the blue vertices. What's more, no two red vertices are adjacent, and no two blue vertices are adjacent. The schematic picture of such a graph is shown in Figure 10. Such graphs are called *bipartite* graphs.

The first two graphs in Figure 11 are familiar. Figure 11(a) is C_8, while Figure 11(b) is the graph N from Figure 5. As far as Figure 11(c) is concerned, it may not be obvious at first that it is bipartite. You may want to check that out. Remember, all you need to do is to show that it's 2-colorable. But one thing also worth noting is that the graph of Figure 11(c) is a tree. This is because it is connected and has no cycles.

If you go back to Figure 10 for a minute, suppose we have a bipartite graph with r red vertices and b blue ones. If it has as many edges as possible, it has b edges coming out of each red vertex and r edges coming out of each blue vertex. We call such a graph a *complete bipartite* graph

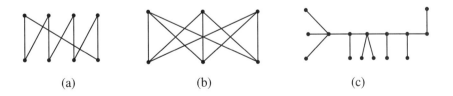

(a)   (b)   (c)

FIGURE 11   Bipartite graphs

with parts of size r and b. We denote that graph by $K_{r,b}$. So the graph of Figure 11(b) is actually $K_{3,3}$, as are the graphs M and N of Figure 5.

● ● ●   **BREAK 11**

    (1) What does a map look like that has a bipartite graph as its dual graph?

    (2) Show that all trees are bipartite graphs.

5.6   ANOTHER CONCEPT

So if we start off with a map, we can produce a graph (the dual graph) from it. And coloring the *faces* of the map gives a coloring of the *vertices* of the graph. Now if all maps are 4-colorable, are all graphs 4-colorable too? Surely this can't be true. If a vertex has 4 neighbors, then won't the graph need 5 colors, one color for the vertex and one for each neighboring vertex? No. We've seen that this isn't the case in Figure 9. The graph there is shown again in Figure 12. Here v has degree 4, but the graph can be colored using only three colors!

Hang on! If all the vertices other than v were joined to each other, then we *would* need more than four colors. Look at the graph in Figure 13.

If v is colored 1, then a could be colored 2. But no other vertex can be colored 1 or 2 because every other vertex is joined both to v and to a. This means that b has to have a new color, 3 say. Arguing in this way we see that each new vertex we look at in Figure 13 has to have a new color. So we need five colors altogether.

Isn't this a difficulty for us? Surely the map we get from the graph in Figure 13 has to have five colors, too. What is that map? How do we get

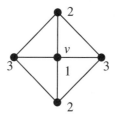

FIGURE 12

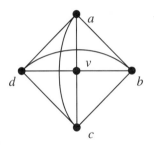

FIGURE 13 A graph that needs 5 colors

it from the graph? Let's try to get the map from the graph and see what it looks like.

The diagram in Figure 14 tells the story. You might as well put V as the middle region, which has to be surrounded by A, B, C, and D. Since A is already touching B and D, it has to be made to touch C. This accounts for the "handle" that goes across between A and C. But then look at B. It's already next to A and C, so how can we make sure that B and D have a common boundary? We obviously need a "handle" between them.

At this stage, there's clearly a problem. The two "handles" have to overlap. So how can we produce a map corresponding to the graph of

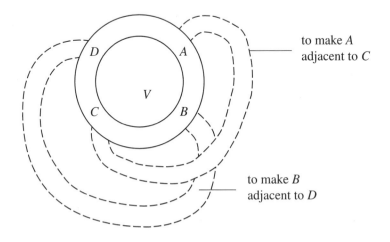

to make A
adjacent to C

to make B
adjacent to D

FIGURE 14 To find the map in the graph of Figure 13

Figure 13? It doesn't look as if we can. Maps can't have regions that cross each other.

This obviously raises the question of which graphs are dual graphs of a map? When they are, we say that the graph has a dual map.

• • • BREAK 12

(1) Are there some graphs that are 6-colorable? Can you find a graph that is n-colorable for any n?
(2) What graphs have dual maps and why?

5.7 PLANARITY

What essential properties do dual graphs have? We know that the degrees of the vertices are not restricted because we can surround a country by as many countries as we like. But we also know that not all graphs can be dual graphs. Figure 13 is an example of a graph that turns out not to have a corresponding map. So how do we know when the graph we have is a dual graph? One very simple property of a dual graph is that, when we draw the graph on top of its corresponding map so that vertex x lies in X, then none of the edges cross. This is because we only join vertices in regions that have a common boundary. We will say that graphs that can be drawn so that no two edges cross are **_planar_**. In Figure 15 we give some examples of planar graphs. Of course, all dual graphs are planar.

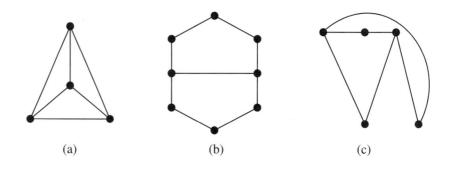

(a)   (b)   (c)

FIGURE 15   Examples of planar graphs

● ● ● **BREAK 13**

(1) If you have a graph drawn so that its edges *do* cross, does that necessarily mean that it is not planar?

(2) The Hiltons, the Holtons, and the Pedersens have bought three neighboring houses. They all want to have gas, electricity, and water connected from the three main points for these utilities. Can the pipes be laid in such a way that no two cross over each other at any point?

Now it's easy to see that dual graphs are planar, but it's another matter to decide, in general, whether or not a given graph is planar. After all, for a graph with a large number of vertices, you might take ages trying to draw it so that no two edges cross. Even if you couldn't find a planar drawing, how would you know that no such drawing existed? You might have missed it somehow.

Let's go back for a moment to the graph in Figure 13. We've drawn it again in Figure 16(a), along with another graph (Figure 16(b)). Now, of course, from what we've said before, Figure 16(a) is the complete graph K_5, and Figure 16(b) is the complete bipartite graph $K_{3,3}$.

Now we have essentially shown that K_5 is non-planar in the discussion relating to Figure 13. Meanwhile, you have considered the planarity of $K_{3,3}$ under another guise in BREAK 13. But we'll now outline an argument that shows that $K_{3,3}$ is not planar.

Let's assume that $K_{3,3}$ is planar. In the final "planar" drawing of $K_{3,3}$ we will have to have the cycle (a, d, b, e, c, f) as shown in Figure 16(c). The question then is, where are the three remaining edges?

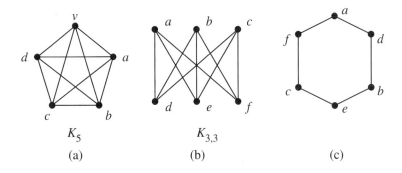

K_5 $K_{3,3}$

(a)   (b)   (c)

FIGURE 16

Now, at the moment, there is a symmetry between the edges *ae*, *bf*, and *cd* and so we can first think about the edge *ae* without any loss of generality. This edge is either drawn *inside* the cycle or *outside*. But if the edge is actually *outside*, we can interchange the inside and outside of the cycle, so we can assume that *ae* is inside. This forces *cd* to be outside the cycle; otherwise, it would have to cross *ae*. This gives us the situation of Figure 17.

The problem is where to put the edge *bf*? If it is *inside* the cycle, it has to cross the edge *ae*. If it is *outside* the cycle, it has to cross the edge *cd*. These both contradict the assumption that $K_{3,3}$ is planar. Hence that assumption is false. So $K_{3,3}$ is not planar!

Discovering that K_5 and $K_{3,3}$ are not planar is no big deal. However, the Polish topologist Kazimierz Kuratowski was able to show that, in a sense, these are the *only* non-planar graphs. Kuratowski's Theorem says that a graph G is non-planar if and only if, in a special sense to be explained, it contains K_5 or $K_{3,3}$ as a subgraph.

To explain to you this special sense, let's see why the graph of Figure 18, the so-called Petersen graph, is non-planar. The highlighted edges in Figure 18 form the subgraph $K_{3,3}$, "in a special sense." Hence, by Kuratowski's Theorem, this graph is non-planar.

Let's think about this special sense. If we take the highlighted subgraph of Figure 18, we get the graph of Figure 19(a). This can be rearranged (see Figure 19(b)) so that it looks more like $K_{3,3}$.

The graph in Figure 19(b) clearly shows $K_{3,3}$ with a small case of measles. To four edges of $K_{3,3}$ have been added vertices of degree 2. Intuitively, it's clear that adding vertices of degree 2 to $K_{3,3}$ cannot suddenly

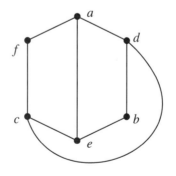

FIGURE 17 $K_{3,3}$ **without the edge** *bf*

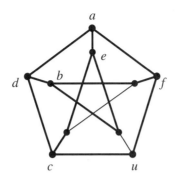

FIGURE 18

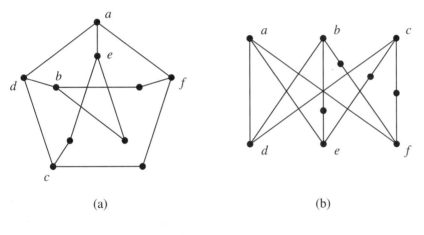

(a)   (b)

FIGURE 19

make it planar. There is no way that adding vertices of degree 2 can *undo* a crossing of edges.

Now the same thing can be said of K_5, too. If K_5 sprouts measles in the form of an arbitrary number of vertices of degree 2, then the resulting graph is still non-planar. Thus Kuratowski's Theorem says that G is non-planar if and only if it contains a subgraph that is either K_5 with measles or $K_{3,3}$ with measles. (Of course, as in much of mathematical language, "with measles" includes the possibility of no measles.)

● ● ●   **BREAK 14**

(1) Show that K_5 is non-planar, using an argument similar to the one we used on $K_{3,3}$.

(2) Show, in three ways, that the graph in Figure 20 is non-planar.

5.8   THE END

We started out with a map. We turned that into a dual graph. We noticed that dual graphs are planar. If we could prove that all planar graphs could be colored with four (or fewer) colors, then, of course, so could all dual graphs. But then all maps would be 4-colorable because dual graphs are. How might we prove that all planar graphs are 4-colorable?

Let's not worry about that for a moment. In 1879, the Englishman Kempe proved that the Four Color Conjecture is true! As a reward for this and other work, Kempe was made a Fellow of the Royal Society. The Royal Society is a body that was set up for the promotion of science in Britain under the Royal Charter of King Charles II in 1662. Only a limited number of Fellows exist at any one time. Existing Fellows elect new Fellows on the basis of their contribution to science. To become an F.R.S. is a singular honor.

Kempe's proof therefore was generally applauded and certainly was widely accepted. However, the story wasn't over. In 1890 another English mathematician, Percy John Heawood, found an error in Kempe's proof! Heawood was, however, able to salvage an important idea, that of Kempe chains (see [6] or [8]), from Kempe's proof. Using Kempe's work, Heawood was able to prove that every map is 5-colorable. (See [2], [3], [4], [6], [8], or [9].)

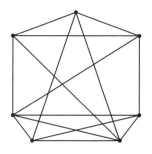

FIGURE 20

For nearly 90 more years mathematicians, both amateur and professional, tried to solve the Four Color Conjecture. Some progress was made. For instance, a group of people concentrated on the number of regions in the map. One mathematician provided his new wife with maps for her to color on their honeymoon! By the mid-1970's it was known that every map with at most 99 regions is 4-colorable.

Then in 1976, Appel and Haken [1] completed the proof using a method strongly based on Kempe's ideas. The major tool that Appel and Haken used that wasn't available to Kempe was a computer. Before they used it, though, they had to exploit certain carefully chosen mathematical reductions to make the problem finite; otherwise, the computer would have been of no use. But, even so, they had to run their machine for 3 months in order to cover all the cases required by their approach.

The nature of this proof raised certain mathematicians' hackles. Up to this time, it had been possible for mathematicians to check every line of a proof and, consequently, to accept the proof or reject it. Here was a proof by computer that, if written out in full, could not have been checked even if all of the world's mathematicians and their apprentices had given the task all the years they had at their disposal. In many quarters this "new" proof took some time to be accepted.

In the meantime, various mathematicians completed computer proofs similar to that of Appel and Haken. To date the nicest attempt at tidying up a computer proof has come from Robertson, Sanders, Seymour, and Thomas (see [8]). But there are still mathematicians who would like an elegant proof without recourse to a computer. A "traditional" proof would make many people happy.

● ● ●   **BREAK 15**

Find, and check, a proof of the Five Color Theorem.

5.9   COLORING EDGES

The Scottish mathematician P.G. Tait had some crazy ideas, but in 1880 he had an idea that might just work. There is still a long way to go, but here is the idea, and here is how things stand up to the publication of this book.

Instead of looking at the dual graph of a map, we'll look at its ***underlying*** graph. In other words, we'll look at the graph whose edges are the boundaries of the regions and whose vertices are the points where three or more boundaries meet.

The first thing to notice is that we may assume that precisely three boundaries meet at every vertex of the underlying graph. We show how to do this in Figure 21. We put a new "polygonal" region at every vertex to give an underlying graph all of whose vertices are of degree 3. We call the corresponding map a ***trivalent map***.

Suppose the trivalent map is 4-colored. Then regions A, B, C, D, E are colored such that A and B have different colors, B and C have different colors, C and D have different colors, D and E have different colors, and E and A have different colors. Hence the 4-coloring of the trivalent map leads to a 4-coloring of the map without region X. This means that we only have to worry about coloring trivalent maps.

So now let's concentrate on trivalent maps. And let's turn to looking at *edge*-coloring their underlying graphs, each vertex of which has degree 3. (By an edge-coloring we mean a coloring of the edges so that edges that meet at a common vertex have different colors.) What we want to show is that 4-coloring the *regions* of the map is equivalent to 3-coloring the *edges* of the underlying graph. This was Tait's idea, and there's a rather nice proof.

Before we can examine the proof, though, we need to restrict the kind of map we are dealing with. In Figure 22 we have a map where the same region lies on two sides of the edge e.

Remove e and the regions on the right. Suppose the left regions plus the outside face are 4-colorable. Remove e and the regions on the left. Suppose the right regions plus the outside face are 4-colorable. Let the outside face in each case have the same color. Then putting the two colorings back together gives a 4-coloring of the original map. The Four Color Conjecture now depends on coloring the left and right submaps. So we'll assume from

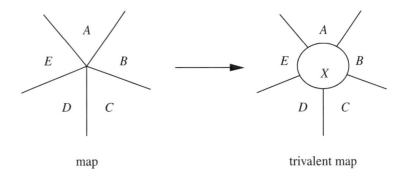

map   trivalent map

FIGURE 21

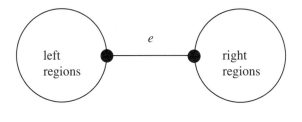

FIGURE 22

now on that we are dealing with the smallest map that is not 4-colorable. Such a map will not have a boundary like e in Figure 22.

To show that a 4-face coloring supplies a 3-edge coloring, we need to take another small detour. We ask for the reader's patience.

Consider the set $W = \{(0, 0), (0, 1), (1, 0), (1, 1)\}$. We'll define the addition of elements of W according to the table below; and we'll write $\oplus$ for this addition.[3]

$\oplus$	$(0, 0)$	$(0, 1)$	$(1, 0)$	$(1, 1)$
$(0, 0)$	$(0, 0)$	$(0, 1)$	$(1, 0)$	$(1, 1)$
$(0, 1)$	$(0, 1)$	$(0, 0)$	$(1, 1)$	$(1, 0)$
$(1, 0)$	$(1, 0)$	$(1, 1)$	$(0, 0)$	$(0, 1)$
$(1, 1)$	$(1, 1)$	$(1, 0)$	$(0, 1)$	$(0, 0)$

Table 1. Addition in W

There are three very useful properties of the set W under the operation $\oplus$. First of all, $c_1 \oplus c_2 = c_2 \oplus c_1$ for all $c_1, c_2 \in W$. You should be able to see, too, that if $c_1 \oplus c_2 = (0, 0)$, then $c_1 = c_2$, and vice versa. Finally, you can check that if $c_1 \oplus c_2 = c_1 \oplus c_3$, then $c_2 = c_3$. We'll find these properties useful in a moment.

Now suppose that we have colored all the faces of a map using four colors represented by the elements of W. We then color the edges of the underlying graph by the sum of the colors on the faces on either side of the edges (see Figure 23).

The thing to note here is that the edge receiving the color $c_1 \oplus c_2$ is not an edge like the edge e in Figure 22. Hence the faces on either side of the edge are different. This means that $c_1 \neq c_2$. From what we said above, we

[3] W is a well-known object in linear algebra, a vector space of dimension 2 over the field with 2 elements, that is to say, the pairs are added componentwise mod 2.

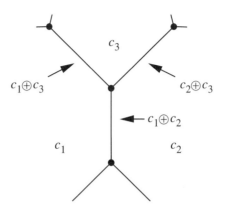

FIGURE 23   Coloring edges using face coloring

now know that $c_1 \oplus c_2 \neq (0, 0)$. This argument is true for *all* edges, and so the colors on the edges of the underlying map must therefore come from the set $\{(0, 1), (1, 0), (1, 1)\}$.

Now, can two adjacent edges have the same color? Suppose $c_1 \oplus c_2 = c_1 \oplus c_3$. From what we know of W, $c_2 = c_3$. Hence we have an edge that has the same face on either side of it. We know that this is not possible. So adjacent edges have different colors. This strange method of coloring edges, given the colors on the faces, has led to a 3-coloring of the edges. The edge colors come from $\{(0, 1), (1, 0), (1, 1)\}$. So we see that a 4-face coloring implies a 3-edge coloring.

How can we go the other way? We now want to 3-edge color the underlying graph of a trivalent map and produce a 4-coloring of its faces. So we can assume that we have the 3-coloring of the edges. Suppose the colors are r, s, and t. The trick here is to pick two colors r and s, and to look first at the edges colored r or s. By themselves, what sort of a graph do they form? Because every vertex of the underlying graph is of degree three, and because we have three colors, all three colors *have* to be present on the three edges around each vertex. So ignoring edges with color t is like throwing those edges away. What that does is to leave us with a graph, all of whose vertices have degree 2. Such a graph has to be a cycle or a collection of cycles. A map consisting of cycles can be face colored with just two colors. Let these two colors be a and b.

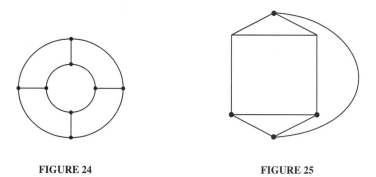

FIGURE 24   FIGURE 25

Now repeat the process by looking just at the edge colors r and t. Again these produce a graph that is a collection of cycles. These can be face colored using colors c and d.

The final trick is to notice that any region in the original map is the intersection of faces from the two lots of cycle maps that we have just described. So color a face ac if it is the intersection of an a-colored face and a c-colored face, or color it ad, bc, bd for the various other intersections. Hence we've used at most four colors. By the method of coloring the cycle graphs, no two faces colored ac, ad, bc, or bd can be adjacent. So a 3-edge coloring gives us a 4-face coloring.

This shows the equivalence of the 4-face coloring problem and the 3-edge coloring problem.

● ● ●   **BREAK 16**

    (1) Show that $c_1 \oplus c_2 = c_2 \oplus c_1$.
        Show that $c_1 \oplus c_2 = (0, 0)$ implies that $c_1 = c_2$.
        Show that $c_1 \oplus c_2 = c_1 \oplus c_3$ implies that $c_2 = c_3$.
    (2) Show that the edges of the map-graph in Figure 24 are 3-edge colorable by first finding a 4-face coloring.
    (3) In the graph of Figure 25 produce a 4-face coloring from a 3-edge coloring.

5.10   A BEGINNING?

Suppose we have an arbitrary planar graph all of whose vertices have degree 3. How can we show that it is 3-edge colorable? If we can find a way to do this, we will have found a way of proving the Four Color

Conjecture. If we can find a method that does not require the use of a computer, that would be a bonus.

Of course, we don't know how to do this . . . yet. But there might be a possible approach hidden in what we do next. Maybe you'll see how to do it. Let us know if and when you do.

There are two key steps in the plan. The first comes from a theorem due to the Russian mathematician V.G. Vizing. He proved that the graphs we are looking at need only three or four colors in order to color their edges! That cuts down the whole business considerably. We don't have to worry about these graphs having to be 5- or 500-edge colored.

Suppose we take one of our planar graphs of degree 3. If it's 3-edge colored, we're finished by what we did in the last section. If it's 4-edge colored . . . ? Ay, there's the rub! So let us assume that it's 4-edge colored. What do 4-edge colored graphs of degree 3 look like? If we knew the answer to that, we might be finished.

Actually, by various tricks we can reduce the kind of 4-edge colored graphs of degree 3 we need to look for. Surprisingly, they turn out to be hard to find. At one stage in the 1970s only five of them were known! Martin Gardner, who wrote a column in *Scientific American* for many years, suggested that searching for these graphs was like hunting for snarks, referring to a poem by Lewis Carroll. Hence the restricted graphs we are looking for came to be known as *snarks*.

The smallest snark is the graph of Figure 18, which we showed to be nonplanar. This graph, as you know, is called the Petersen graph; it is named after the Scandinavian mathematician Julius Peter Christian Petersen, who did some early work with it. We show the Petersen graph again in Figure 26.

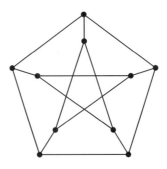

FIGURE 26   **The Petersen graph**

• • • **BREAK 17**

Show that the Petersen graph is 4-edge colorable but not 3-edge colorable.

Perhaps by giving snarks some publicity in *Scientific American*, Gardner prompted the mathematical community into action. Relatively soon after his column appeared, more snarks were discovered. In fact, it was soon established that there is an *infinite* number of them. This was done by showing how to construct new snarks from old ones. We give such a construction below.

This construction is called the ***dot product*** of two graphs and was produced by R. Isaacs [5] in 1975. He first took two snarks, any two snarks, L and R. Then he

1. removed a pair of adjacent vertices x and y from L;
2. removed a pair of independent edges ab and cd from R;
3. joined vertices between L and R by means of new edges ra, sb, tc, ud, as shown in Figure 27, to produce a graph, $L.R$, which is regular of degree 3.

It turns out that $L.R$ is a snark.

In Figure 27 we have chosen to join r to a, s to b, t to c, and u to d, but we would have obtained a snark, which we would still refer to as $L.R$, if instead of joining the vertices of $\{r, s\}$ to the vertices of $\{a, b\}$ and the vertices of $\{t, u\}$ to the vertices of $\{c, d\}$ we had joined the vertices of $\{r, s\}$ to the vertices of $\{c, d\}$, and the vertices of $\{t, u\}$ to the vertices of $\{a, b\}$.

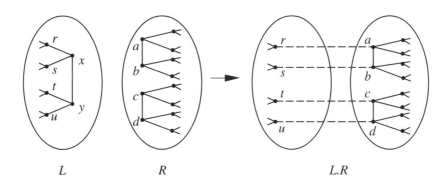

FIGURE 27

To see how this works out in practice, let's construct an example of $P.P$, where P is the Petersen graph. The graph shown in Figure 28 is one of the Blanuša snarks, named after the Croatian mathematician Danila Blanuša. This graph is one of the five first-known snarks we mentioned above.

Using P alone as the starting snark we can produce an infinite collection of snarks via $P.P$, $P.(P.P)$, $P.(P.(P.P))$, $\cdots$. But these are not the only snarks known. For many more examples see [7].

• • • BREAK 18

Find another snark of the form $P.P$ that is not isomorphic to the graph in Figure 28.

The surprising thing about all the snarks that we know today is that, in some sense, they *all* contain the Petersen graph! Here we have to be vague about what "in some sense" means. Roughly, it means that the Petersen graph is a subgraph of them all.

This led the British mathematician W.T. (Bill) Tutte, who spent much of his working life at Waterloo University in Canada, to make the following conjecture.

Tutte's Conjecture   *Every snark contains a copy of the Petersen graph.*

This is our second key step. If this is true, then our imagined 4-edge colorable, planar graph of degree 3 has the Petersen graph inside it somewhere. If it does, then, by Figure 18, the graph contains a copy of $K_{3,3}$. Hence the graph is non-planar. But this contradicts the assumption that

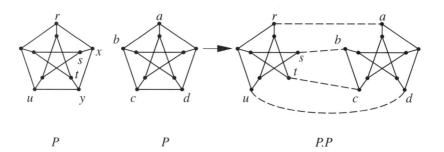

$$P \hspace{4cm} P \hspace{4cm} P.P$$

FIGURE 28

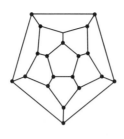

FIGURE 29

it was *planar*! Hence it can't be 4-edge colorable. And we've proved the Four-Color Theorem by another route!

● ● ● **BREAK 19**

(1) Find all graphs on n vertices that are regular of degree 1.

(2) Find all graphs on n vertices that are regular of degree 2.

(3) Find all graphs on n vertices that are regular of degree 3. (If this is too hard, try it for $n = 4$ and $n = 6$.)

(4) Show that the "dodecahedron"[4] in Figure 29 is **Hamiltonian**. That is, show that there is a cycle in the graph that passes through all the vertices of the graph.

(5) Use Kuratowski's theorem to show that K_n is non-planar for all $n \geq 5$. Which complete bipartite graphs are planar?

(6) Find all connected graphs on 5 vertices that have no cycles.

(7) What is the smallest number of colors that can be used to edge-color $K_{3,3}$?

(8) Show that if every planar graph that is regular of degree 3 were Hamiltonian, then the Four-Color Theorem would follow. (See question (4) for a definition of Hamiltonian.)

REFERENCES

1. Appel, K. and W. Haken, The solution of the four-color-map problem. *Scientific American*, 237(4), 1977, 108–121.

2. Beineke, L.W. and R.J. Wilson (eds), *Selected Topics in Graph Theory* 1, 2, Academic Press, London, 1978, 1983.

---

[4]The figure shows the edges of a regular dodecahedron as viewed up close through one face. This is called the **Schlegel diagram** of the dodecahedron.

3. Biggs, N.L., E.K. Lloyd, and R.J. Wilson, *Graph Theory 1736–1936*, Clarendon Press, Oxford, 1986.
4. Clark, J. and D.A. Holton, *A First Look at Graph Theory*, World Scientific, Singapore, 1996.
5. Isaacs, R., Infinite families of non-trivial trivalent graphs which are not Tait colorable, *Amer. Math. Monthly*, **82**, 1975, 221–239.
6. Fritsch, R and G. Fritsch, *The Four-Color Theorem*, translated from the German by J. Peschke, Springer Verlag, New York, 1998.
7. Holton, D.A. and J. Sheehan, *The Petersen Graph*, Cambridge University Press, Cambridge, 1987.
8. Robertson, N., D.P. Sanders, P.D. Seymour, and R. Thomas, The Four-Color Theorem, *J. Comb. Th.* (B), **70**, 1997, 2–44.
9. Saaty, T.L. and P.C. Kainen, *The Four-Color Problem*, McGraw-Hill, New York, 1977.
10. Wilson, R.J., *Introduction to Graph Theory*, Third Edition, Longman, Harlow, England, 1983.

Information on the web regarding The Four Color Theorem can be found at `http://www.astro.virginia.edu/~eww6n/math/Four-ColorTheorem.html` and at `http://www.math.gatech.edu/~thomas/FC/fourcolor.html`

6

From Binomial to Trinomial Coefficients and Beyond

6.1 INTRODUCTION AND WARM-UP

In [3 (Chapter 6, Part I), 5, and 6] we presented the binomial coefficients $\binom{n}{r}$ in a geometric, an algebraic, and a combinatorial framework, being much concerned with establishing interesting connections between their algebraic properties and geometric features of the Pascal Triangle.

We also introduced in [3], but did not exploit to its fullest, a far more revealing notation for a binomial coefficient than $\binom{n}{r}$, namely, $\binom{n}{r\ s}$, where[1] $r + s = n$. This notation emphasizes the *symmetry* of the binomial coefficient, so that we have the **Symmetry Identity**, expressed in symmetric form as

$$\binom{n}{r\ s} = \binom{n}{s\ r} \tag{1}$$

(see Figures 1, 2).

---

[1] The eminent French mathematician Henri Cartan goes one step further than us down the road of notational innovation and writes (r, s) instead of $\binom{n}{r\ s}$. Of course, we are not recommending abandoning the notation $\binom{n}{r}$.

It also brings out the *symmetric* nature of the **Pascal Identity**

$$\binom{n}{r\ \ s} = \binom{n-1}{r-1\ \ s} + \binom{n-1}{r\ \ \ s-1} \tag{2}$$

Notice how much less revealing (1) and (2) would be if s were suppressed. Notice, too, that in Figure 1(a) the Pascal Identity may be expressed by saying that the sum of the two numbers in the hexagons that touch the

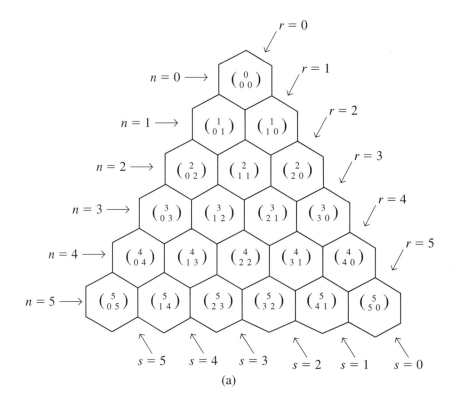

(a)

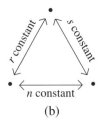

(b)

FIGURE 1   Note that $\binom{n}{r\ s}, r+s=n$, is just the usual $\binom{n}{r}$.

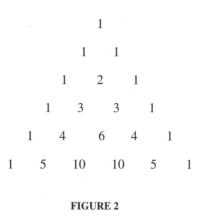

FIGURE 2

"roof" of any hexagon of entries equals the number in the center of that hexagon.[2] Of course, the hexagons may be continued to the right and left of the array in Figure 1(a), but all of these further hexagons would have 0's in their centers (see the Vanishing Identity (7)). Figure 2 gives the numerical value of $\left(\begin{smallmatrix} n \\ r\ s \end{smallmatrix}\right)$ in the part of the Pascal Triangle corresponding to Figure 1(a).

In this chapter we will be using the more explicit notation almost exclusively,[3] and we begin by recalling some of the facts developed in [3], presenting them in this notation.

From the combinatorial interpretation of the binomial coefficient, that is, as the number of ways of choosing r objects from n objects, we see that

$$\binom{n}{r\ \ s} = \frac{n(n-1)(n-2)\cdots(n-r+1)}{r!} \tag{3}$$

We also adopt the natural convention

$$\binom{n}{0\ \ n} = \binom{n}{n\ \ 0} = 1 \tag{4}$$

Remembering that $r + s = n$, we see that (3) may be written neatly, by multiplying the top and bottom by $s!$, as

$$\binom{n}{r\ \ s} = \frac{n!}{r!s!}, \qquad \text{if} \quad n \geq 0, \quad r \geq 0, \quad s \geq 0 \tag{5}$$

---

[2]We talk of the roof of a hexagon to prepare for the analogy with the Pascal Identity for trinomial coefficients (compare Figure 14(c)).

[3]We do this in order to familiarize the reader with the new notation, not because we believe (in fact we don't!) that it always confers an advantage. You will find that, as a general rule, we do not use this expanded notation in Chapter 7.

Notice that, although we have lost the symmetry in (3), we compensate by having a formula that is far more suitable for computation than (5), especially if we avail ourselves of the symmetry property (1) at the same time.

We call (5) the **_arithmetical_** interpretation of $\binom{n}{r\ s}$. Of course, (5) is valid for either $r = 0$ (and $s = n$) or $s = 0$ (and $r = n$), provided that we interpret 0! in the standard way as 1. For binomial coefficients we would not, as we have said, normally use (5) to calculate their value, but, as you will see, when we calculate multinomial coefficients — in particular trinomial coefficients — it is natural to rely on a form similar to (5), since we do not have a convenient form generalizing (3).

A second interpretation of the binomial coefficient is obtained by observing that **_algebraically_** $\binom{n}{r\ s}$ may be thought of as _the coefficient of $a^r b^s$ in the expansion of $(a + b)^n$, with $r + s = n$_. Thus

$$(a + b)^n = \sum_{r+s=n} \binom{n}{r\ s} a^r b^s \tag{6}$$

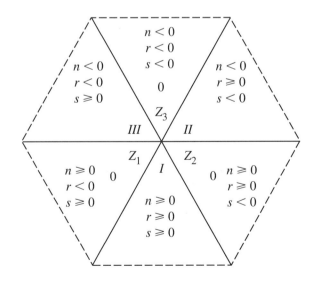

FIGURE 3   **Remember, $r + s = n$.**

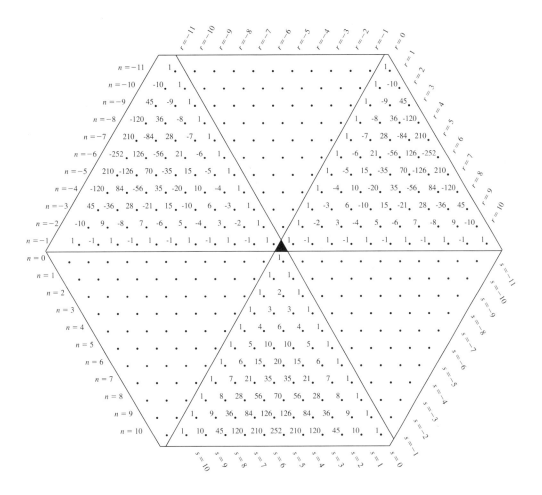

FIGURE 4   The Pascal Hexagon for $-11 \leq n, r, s \leq 10$. Note that we have slightly displaced the frontier lines between sextants in order to be able to display the non-zero values along these lines.

From (6) we immediately obtain the[4] *Vanishing Identity*

$$\binom{n}{r \quad s} = 0 \quad \text{for} \quad n \geq 0 \quad \text{and} \quad r < 0 \quad \text{or} \quad s < 0 \quad (7)$$

which explains the two lower zero triangular sextants in the Pascal hexagon in Figures 3 and 4. (In Figure 4 each unadorned dot stands for a 0 entry.)

[4] Although it would be equivalent to state the conditions on (7) as "for $n \geq 0$ and $r > n$ or $s > n$," this would not suggest the analogous conditions for trinomial coefficients.

Before considering what happens when $n < 0$, let us take a break and look at some patterns in the Pascal triangle.

• • • BREAK 1

(1) Make yourself a Pascal Triangle with $0 \leq n \leq 10$ and, for the following identity, circle the values involved in the identity for $n = 3, k = 5$:

$$\sum_{i=0}^{k} \binom{n+i}{n \quad i} = \binom{n+k+1}{n+1 \quad k} \tag{8}$$

You may then see why (8) is sometimes called the **Christmas Stocking Theorem** (with the stocking hanging on one of the 1's along the right-hand side of the Pascal Triangle).

By symmetry you will then see that the statement of the Christmas Stocking Theorem where the stocking hangs on one of the 1's along the left-hand side of the Pascal Triangle is

$$\sum_{i=0}^{k} \binom{n+i}{i \quad n} = \binom{n+k+1}{k \quad n+1} \tag{9}$$

(2) Prove either (8) or (9). (Hint: Use the Pascal Identity to subtract $\binom{n+k}{k \ n}$ from $\binom{n+k+1}{k \ n+1}$. What happens?) Use the same type of argument to prove that

$$\binom{n}{r-1 \quad s+1} + \sum_{i=0}^{k} \binom{n+i}{r+i \quad s} = \binom{n+k+1}{r+k \quad s+1}$$

[Hint: Notice that this coincides with (9) above if $s = n$, by the Vanishing Identity.]

(3) A few years ago, one of the authors (DH) was working with a child with autism who enjoyed playing with baked-bean and spaghetti cans.[5] In an effort to use this interest to understand what mathematics the child could do, DH brought him the cans to work with. The first thing the child became interested in doing with the cans was making towers. Eventually, no doubt with some prodding from DH, he tried to see how many different kinds of towers he could make with 2 cans of baked beans and 2 cans

---

[5] We even published a paper on this whole story; see [1].

of spaghetti available. After a long time experimenting the child
displayed his answer this way:

$$
\begin{array}{ccccccccc}
 & & & & & B & S & S \\
 & & S & B & S & B & S & B & S \\
S & B & S & S & B & B & S & S & B \\
\\
 & & & & B & S & B & S & S & B \\
 & & S & B & B & B & B & S & S & B & S \\
 & & B & S & B & S & B & B & B & S & S \\
 & & B & B & S & S & S & S & B & B & B
\end{array}
$$

If we include the tower of height 0, calling it the empty tower E,
then we get a total of 19 towers. These 19 towers may be placed
more suggestively as follows:

$$
\begin{array}{ccccc}
 & & \underbrace{E} & & \\
 & \underbrace{B} & & \underbrace{S} & \\
B & & B \quad S & & S \\
\underbrace{B} & & \underbrace{S \quad B} & & \underbrace{S} \\
 & S \; B \; B & & B \; S \; S & \\
 & B \; S \; B & & S \; B \; S & \\
 & \underbrace{B \; B \; S} & & \underbrace{S \; S \; B} & \\
 & & B \; S \; B \; S \; S \; B & & \\
 & & B \; B \; S \; S \; B \; S & & \\
 & & S \; B \; B \; B \; S \; S & & \\
 & & \underbrace{S \; S \; S \; B \; B \; B} & &
\end{array}
$$

Notice how many towers are atop each brace in this diagram
and how these numbers are related to the binomial coefficients.
Use (8) and then (9) to show that the sum of all these numbers is
just $\binom{6}{3\ 3} - 1$. Figure 5 may suggest how to do this. The underlined
entry in Figure 5(a) is the result of adding all the entries along
a line inside the parallelogram (including the endpoints) where
r is fixed. The doubly-underlined entry in Figure 5(b) is the

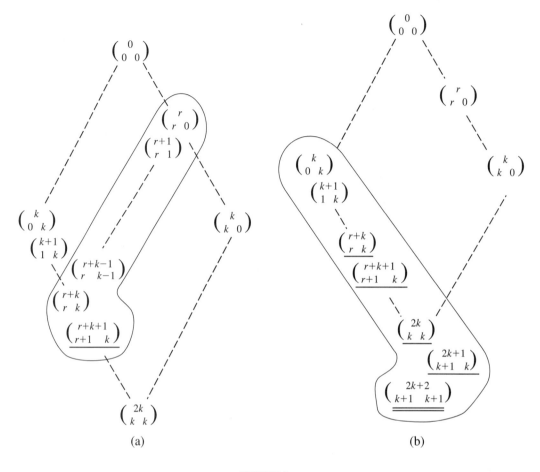

(a)   (b)

FIGURE 5

result of adding all the underlined entries on the boundary of the
parallelogram where s is fixed at $s = k$, starting with the entry
$\binom{k}{0\ k}$. Notice that this means that $\binom{k}{0\ k}$ gets added twice.

(4) Show, by first applying (8) as shown in Figure 5(a), then applying (9) as shown in Figure 5(b), that the number of towers you
can construct, including the empty tower, with k baked beans
cans and k spaghetti cans is $\binom{2(k+1)}{k+1\ \ k+1} - 1$.

(5) Use the same reasoning to show that the number of towers you
can construct, including the empty tower, with k baked bean cans
and ℓ spaghetti cans available is $\binom{k+\ell+2}{k+1\ \ \ell+1} - 1$ (see Figure 6).

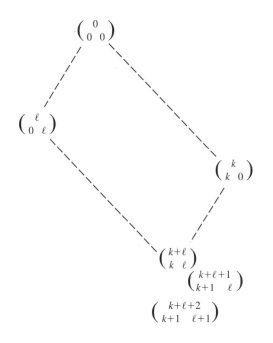

FIGURE 6

(6) Use the result of part (5) to show that the sum of the binomial coefficients on the boundary and interior of the parallelogram having vertices located at the binomial coefficients (see Figure 7)

$$\begin{pmatrix} n \\ r \ \ s \end{pmatrix}, \quad \begin{pmatrix} n+\ell \\ r \ \ \ s+\ell \end{pmatrix}, \quad \begin{pmatrix} n+k \\ r+k \ \ s \end{pmatrix}, \quad \begin{pmatrix} n+k+\ell \\ r+k \ \ \ s+\ell \end{pmatrix}$$

is

$$\begin{pmatrix} n+k+\ell+2 \\ r+k+1 \ \ \ s+\ell+1 \end{pmatrix} - \begin{pmatrix} n+k+1 \\ r+k+1 \ \ \ s \end{pmatrix}$$

$$- \begin{pmatrix} n+\ell+1 \\ r \ \ \ s+\ell+1 \end{pmatrix} + \begin{pmatrix} n \\ r \ \ s \end{pmatrix}$$

We now return to our extension of the Pascal Triangle, already started with the vanishing identity (7), and ask what happens when n is a negative integer. There are then, in fact, two possible power series expansions of

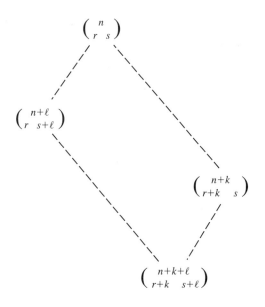

FIGURE 7

$(a + b)^n$, one valid if $|a| < |b|$ and the other valid if $|a| > |b|$. By Taylor series arguments[6] we obtain, with $n < 0$, (see [4, 7]),

$$
\begin{cases}
\text{if} \quad |a| < |b|, \quad \text{then} \quad (a + b)^n = \displaystyle\sum_{r \geq 0} \binom{n}{r \quad s} a^r b^s \\[4mm]
\text{where} \quad \binom{n}{r \quad s} \stackrel{\text{def}}{=} (-1)^r \binom{-s-1}{r \quad -n-1}
\end{cases}
\tag{10}
$$

and, by symmetry,

$$
\begin{cases}
\text{if} \quad |a| > |b|, \quad \text{then} \quad (a + b)^n = \displaystyle\sum_{s \geq 0} \binom{n}{r \quad s} a^r b^s \\[4mm]
\text{where} \quad \binom{n}{r \quad s} \stackrel{\text{def}}{=} (-1)^s \binom{-r-1}{-n-1 \quad s}
\end{cases}
\tag{11}
$$

[6]We might also use the convergent infinite series $1 + x + x^2 + \cdots = \frac{1}{1-x}$, valid if $|x| < 1$, and differentiate $(-n - 1)$ times (provided that we can justify the term-by-term differentiation!).

Thus we are led to define

$$\binom{n}{r\ \ s} = \begin{cases} \dfrac{n!}{r!s!}, & n \geq 0, \quad r \geq 0, \quad s \geq 0 \\[2mm] (-1)^r \dbinom{-s-1}{r\ \ -n-1}, & n < 0, \quad r \geq 0, \quad s < 0 \\[2mm] (-1)^s \dbinom{-r-1}{-n-1\ \ s}, & n < 0, \quad r < 0, \quad s \geq 0 \end{cases} \qquad (12)$$

We complete the Pascal Hexagon in the obvious way by setting

$$\binom{n}{r\ \ s} = 0 \qquad \text{if} \quad n < 0, \quad r < 0, \quad s < 0 \qquad (13)$$

We note that, with $\binom{n}{r\ s}$ so defined, we have extended (1) and (2) so that they are now valid for all integers n.

Symmetry Identity

$$\binom{n}{r\ \ s} = \binom{n}{s\ \ r} \qquad (14)$$

Pascal Identity

$$\binom{n}{r\ \ s} = \binom{n-1}{r-1\ \ s} + \binom{n-1}{r\ \ s-1} \qquad (15)$$

  except that

$$\binom{0}{0\ \ 0} \neq \binom{-1}{-1\ \ 0} + \binom{-1}{0\ \ -1} \qquad (16)$$

However, if you examine (10) and (11) you will see that we should not expect the Pascal Identity to be valid in this one exceptional case, as the binomial coefficients $\binom{-1}{-1\ 0}$ and $\binom{-1}{0\ -1}$ arise in conflicting situations. Thus (16) serves as a warning rather than an exception!

Certain qualitative features of the Pascal Hexagon are shown in Figure 3. Notice that the directions in which n, r, and s are constant are just as shown in Figure 1(b). Also observe the natural division of the hexagon into six triangular sextants. Three of these are *non-zero sextants*, or *blades*, labeled *I, II, III*, in which *every* entry is non-zero; and three are *zero* sextants, labeled Z_1, Z_2, Z_3, in which *every* entry is zero.[7] Note that the frontier lines between sextants are all regarded as belonging to the *non-zero* blades they abut, except that the center point belongs to *I*.

Notice, in Figure 4, how the absolute values of the numerical values of the entries in the northwest and northeast blades of the Windmill are simply the values in the southern blade rotated through $\pm 120°$. Observe, too, how the signs alternate in successive rows of r, s in the northeast and northwest blades, respectively.

All these facts were presented in [3], but their symmetric aspect may not have caught the reader's attention because we did not always insert the s in the symbol for the binomial coefficient.

In this chapter we are concerned in the first instance with the natural extension of the ideas on binomial coefficients presented in [3] — for example, the generalized Star of David Theorem (which we will discuss in the next break) — from binomial to trinomial coefficients. Once again, our new notation is admirably suited to the passage from 2 dimensions to 3 dimensions. Specifically, a trinomial coefficient $\left(\begin{smallmatrix} & n & \\ r & s & t \end{smallmatrix}\right)$, $r + s + t = n$, is first defined for n a non-negative integer. Then r, s, t are also non-negative integers, and we interpret the trinomial coefficient $\left(\begin{smallmatrix} & n & \\ r & s & t \end{smallmatrix}\right)$ algebraically as the coefficient of $a^r b^s c^t$ in the *trinomial* expansion of $(a + b + c)^n$. It is then easy to show that, in this *positive sector*, which we naturally call the Pascal Tetrahedron, the value of the trinomial coefficient is given by

$$\begin{pmatrix} & n & \\ r & s & t \end{pmatrix} = \frac{n!}{r!s!t!} \tag{17}$$

Of course, the *symmetry* of the trinomial coefficient is now expressed by the fact that the value of $\left(\begin{smallmatrix} & n & \\ r & s & t \end{smallmatrix}\right)$ is unchanged by *any* permutation of r, s, t (there are 6 such permutations, including the identity permutation); and the Pascal Identity, in the case $r, s, t \geq 1$, takes the form

$$\begin{pmatrix} & n & \\ r & s & t \end{pmatrix} = \begin{pmatrix} & n-1 & \\ r-1 & s & t \end{pmatrix} + \begin{pmatrix} & n-1 & \\ r & s-1 & t \end{pmatrix} + \begin{pmatrix} & n & \\ r & s & t-1 \end{pmatrix} \tag{18}$$

Just like the binomial coefficient, the trinomial coefficient also has a *combinatorial* interpretation. It is then to be understood as the number of ways of partitioning a set of n objects into 3 disjoint subsets, of r, s, t objects, respectively.

Since we, and our readers, would wish to be able to generalize beyond trinomial coefficients to *multinomial* coefficients, we introduce in Section 2

---

[7]We often refer to the **Pascal Windmill** if we want to concentrate attention on the 3 non-zero triangular sextants of the Pascal Hexagon. These sextants may then be called **blades**.

a notation suitable for such a generalization. Thus we use, as an alternative to the notation $\left(\begin{smallmatrix} & n & \\ r & s & t \end{smallmatrix}\right)$, the notation $\left(\begin{smallmatrix} & n & \\ k_1 & k_2 & k_3 \end{smallmatrix}\right)$; after all, everybody knows how to write the mth term of the sequence $(k_1, k_2, k_3, \cdots)$, but nobody knows how to write the mth term of the sequence $(r, s, t, \cdots)$! This gives us the opportunity to make the point that different notations have different advantages in different contexts.

The actual content of Section 2 is concerned with the effect of sliding parallelepipeds, the exact analogy of the sliding of parallelograms in [3, Section 6.4], and we deduce, as in that reference, a (generalized) Star of David Theorem. However, this 3-dimensional version is far richer than the 2-dimensional original because of the wider scope of the symmetry concept in 3 dimensions, as indicated above. Thus (compare (1))

$$\begin{pmatrix} & n & \\ r & s & t \end{pmatrix} = \begin{pmatrix} & n & \\ r & t & s \end{pmatrix} = \begin{pmatrix} & n & \\ s & r & t \end{pmatrix} = \begin{pmatrix} & n & \\ s & t & r \end{pmatrix} = \begin{pmatrix} & n & \\ t & r & s \end{pmatrix} = \begin{pmatrix} & n & \\ t & s & r \end{pmatrix}$$

$$(19)$$

If you are unfamiliar with the idea of a symmetry group, we recommend you, at this point, to glance at the material on *symmetry groups* in Chapter 8, especially the definition of the symmetric group S_n, to see that it is the symmetric group S_3, with $3! = 6$ elements, that acts on the triple (r, s, t) and under which, as (19) states, $\left(\begin{smallmatrix} & n & \\ r & s & t \end{smallmatrix}\right)$ is invariant.

In Section 3 we will extend the definition of $\left(\begin{smallmatrix} & n & \\ r & s & t \end{smallmatrix}\right)$ to allow n, r, s, t to take as values any integers — positive, zero, or negative — subject only to the condition $r + s + t = n$. We will also show how to generalize this development to multinomial coefficients $\left(\begin{smallmatrix} & & n & \\ k_1 & k_2 & \cdots & k_m \end{smallmatrix}\right)$. In Section 4 we will discuss variants and other generalizations of our theorems in Section 2 and then, in Section 5, we explore the geometry of the 3-dimensional analog of the Pascal Hexagon.

However, two final, related points should be made. First, when we talk, in Section 2, of the edges of the parallelepiped $P^{(n)}$ having lengths a_1, a_2, a_3, we are referring not to a metric valid over the whole of three-dimensional space, but to a distance measured along the lines $k_2 = $ constant, $k_3 = $ constant (i.e., measured by the change in k_1), or to a distance measured along the lines $k_1 = $ constant, $k_3 = $ constant (i.e., measured by the change in k_2); or to a distance measured along the lines $k_1 = $ constant, $k_2 = $ constant (i.e., measured by the change in k_3).[8] Remember that, from the strict point of view of coordinate geometry, we are

---

[8]Notice our use here of k_1, k_2, k_3 rather than r, s, t. The point we are making readily generalizes to m dimensions and is not at all confined to the case $m = 3$. Indeed, a similar remark would appropriately have accompanied Chapter 6 of [3], with $m = 2$.

looking at the hyperplane $n = k_1 + k_2 + k_3$ in 4-dimensional space — but this pedantic interpretation is not helpful to our algebraic purposes.

Second, in keeping with the convention explained above, we have often drawn, e.g., in Figure 15, a rectangular box instead of a more general parallelepiped. We have felt justified in stylizing in this way, since the angles of the parallelepiped play no role in the argument and the lengths we refer to are merely distances measured along the given edges. Nevertheless, our colleague Hans Walser was kind enough to provide figures which more truly reflect the positions of the parallelepipeds in question in the hyperplane $n = k_1 + k_2 + k_3$ (or $n = r + s + t$) in 4-dimensional space! These lovely diagrams are Figures 15A, 16A, 17A.

• • • **BREAK 2**

(This break is designed to prepare you for Section 2.)

Suppose you have a parallelogram whose vertices are entries in the Pascal Triangle and whose sides run in two of the three possible directions $n = $ constant; $r = $ constant; $s = $ constant. Suppose, further, that the lengths of the sides of this parallelogram are k and ℓ (see Figure 8). Suppose also that each parallelogram has an **anchor** $\otimes$ given by the binomial coefficient $\binom{n}{r \ \ s}$, and define the **cross-product**, or **weight**, of the parallelogram to be the product of the anchor and the

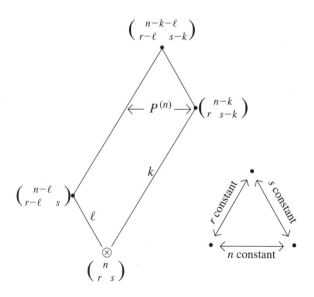

FIGURE 8 A fundamental parallelogram $P^{(n)}$ with the sliding direction shown by arrows.

binomial coefficient opposite it, divided by the product of the other two binomial coefficients (at the ends of the shorter diagonal) of the parallelogram.

Thus the parallelogram $P^{(n)}$ shown in Figure 8 has sides running in the directions where r and s are constant, and its weight $W(P^{(n)})$ is given by[9]

$$W\left(P^{(n)}\right) = \frac{\begin{pmatrix} n \\ r \ \ s \end{pmatrix}\begin{pmatrix} n-k-\ell \\ r-\ell \ \ s-k \end{pmatrix}}{\begin{pmatrix} n-k \\ r \ \ s-k \end{pmatrix}\begin{pmatrix} n-\ell \\ r-\ell \ \ s \end{pmatrix}}$$

where all binomial coefficients on the right must, of course, have all entries non-negative.

(1) Simplify $W(P^{(n)})$ to show that, for fixed k and ℓ, $W(P^{(n)})$ is independent of r and s and thus does not change as we move $P^{(n)}$ in the horizontal direction (i.e., the direction in which n remains constant).[10]

(2) Observe, from what you did in part (1), that $W(P^{(n)})$ is invariant under the interchange of k and ℓ. Thus, in Figure 9, $W(\overset{*}{P}{}^{(n)}) = W(P^{(n)})$. Use this fact to show that

$$\begin{pmatrix} n-k-\ell \\ r-\ell \ \ s-k \end{pmatrix}\begin{pmatrix} n-k \\ r-k \ \ s \end{pmatrix}\begin{pmatrix} n-\ell \\ r \ \ s-\ell \end{pmatrix}$$

$$= \begin{pmatrix} n-k-\ell \\ r-k \ \ s-\ell \end{pmatrix}\begin{pmatrix} n-\ell \\ r-\ell \ \ s \end{pmatrix}\begin{pmatrix} n-k \\ r \ \ s-k \end{pmatrix} \qquad (20)$$

(3) Identify the 6 binomial coefficients appearing in (20) in Figure 10. Describe a *geometric* way of stating (20).

(4) Repeat (with the obvious changes) part (1) for the parallelograms $P_{(s)}$, $P_{(r)}$ shown in Figure 11 and hence establish the three *Star of David*

---

[9]We can, of course replace k or ℓ by $-k$ or $-\ell$, respectively. When both are replaced the weight takes the form

$$W\left(P^{(n)}\right) = \frac{\begin{pmatrix} n \\ r \ \ s \end{pmatrix}\begin{pmatrix} n+k+\ell \\ r+\ell \ \ s+k \end{pmatrix}}{\begin{pmatrix} n+k \\ r \ \ s+k \end{pmatrix}\begin{pmatrix} n+\ell \\ r+\ell \ \ s \end{pmatrix}},$$

which simplifies easily to

$$\frac{n!(n+k+\ell)!}{(n+k)!(n+\ell)!}.$$

[10]This explains why we call this parallelogram $P^{(n)}$.

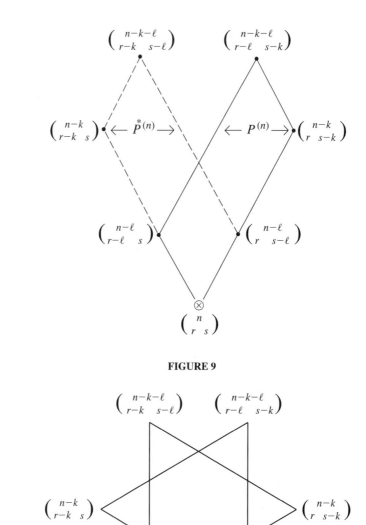

FIGURE 9

FIGURE 10

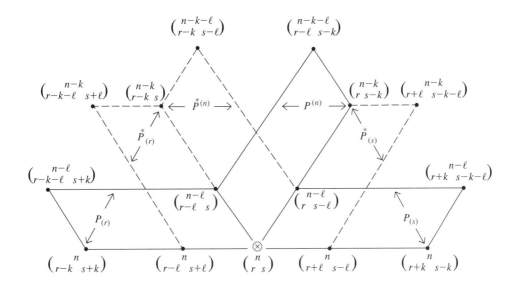

FIGURE 11   The Pascal Flower.

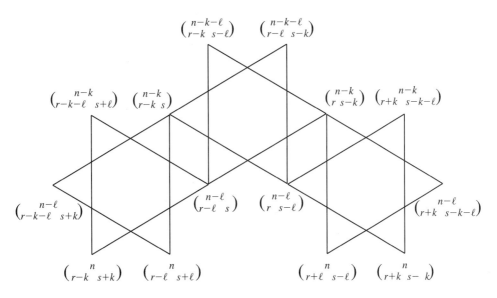

FIGURE 12   The three Stars of David (our picture shows genuine stars ($k = 2\ell$) but that isn't necessary for the argument).

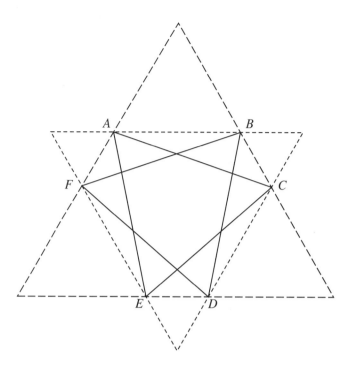

FIGURE 13 Showing how the stars of Figure 12 (even when $k \neq 2\ell$) could have come from either of two equilateral triangles, one pointing up and the other pointing down. Notice that when $k \neq 2\ell$ the original triangle, and the triangles cut off, are each of different sizes in the two cases.

theorems shown in Figure 12. Notice in the process that $W(P_{(s)})$ does not change as we move the vertices of P in the direction in which s remains constant;[11] and similarly that $W(P_{(r)})$ does not change as we move the vertices of P in the direction in which r remains constant.

(5) Compare the expressions for the three weights $W(P^{(n)})$, $W(P_{(s)})$, and $W(P_{(r)})$, noting the symmetry.

(6) Choose, in *any* non-zero blade of the Pascal Windmill, three binomial coefficients that form an equilateral triangle (pointing either up or down). Then cut off from that triangle three identical smaller equilateral triangles, one from each vertex, so that you obtain a semi-regular

---

[11] Notice here that we have placed the "(s)" as a subscript rather than a superscript. This is because s plays a different role from n in the binomial coefficient — as is reflected by its different position in the symbol $\binom{n}{r}_s$.

hexagon with binomial coefficients located at points A, B, C, D, E, F, as shown in Figure 13. Then verify that, in your example,

$$A \times C \times E = B \times D \times F.$$

6.2 ANALOGUES OF THE GENERALIZED STAR OF DAVID THEOREMS

First we define the trinomial coefficient as *the coefficient of $a^r b^s c^t$ in the expansion of $(a + b + c)^n$, with $r + s + t = n \geq 0$*. Thus

$$(a + b + c)^n = \sum_{r+s+t=n} \begin{pmatrix} n \\ r \ s \ t \end{pmatrix} a^r b^s c^t \tag{21}$$

From (21) we immediately obtain the

Vanishing Identity

$$\begin{pmatrix} n \\ r \ s \ t \end{pmatrix} = 0 \quad for \quad n \geq 0 \quad and \quad r < 0 \quad or \quad s < 0 \quad or \quad t < 0 \tag{22}$$

(compare with (7)), which explains why the non-zero trinomial coefficients form triangular layers in the Pascal Tetrahedron (see Figures 14(a), (b) and (c)).

Now the coefficient $\begin{pmatrix} n \\ r \ s \ t \end{pmatrix}$ in (21) above may be thought of as the number of ways of choosing r of the n factors from which you select a, namely, $\begin{pmatrix} n \\ r \ s+t \end{pmatrix}$; and then choosing s of the remaining $s + t$ factors from which you select b, namely, $\begin{pmatrix} s+t \\ s \ t \end{pmatrix}$. Thus, we see that, multiplying these binomial coefficients together, we have

$$\begin{pmatrix} n \\ r \ s \ t \end{pmatrix} = \begin{pmatrix} n \\ r \ s+t \end{pmatrix} \begin{pmatrix} s+t \\ s \ t \end{pmatrix},$$

and this simplifies to

$$\begin{pmatrix} n \\ r \ s \ t \end{pmatrix} = \frac{n!}{r!(s+t)!} \cdot \frac{(s+t)!}{s!t!} = \frac{n!}{r!s!t!} \tag{23}$$

From (23), or from the algebraic or combinatorial[12] interpretation, it is obvious that we have the

---

[12] Recall the combinatorial interpretation following (12).

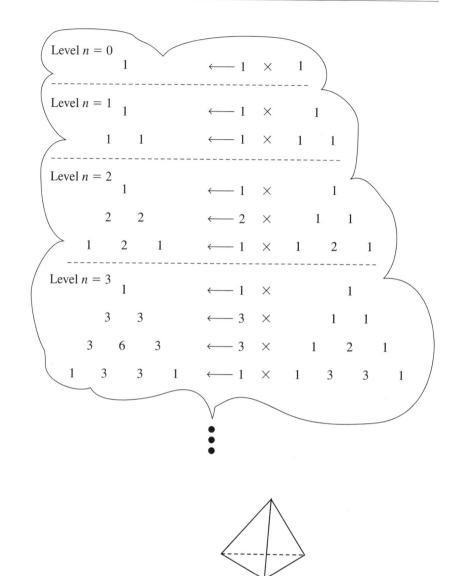

(a)

FIGURE 14   The Pascal Tetrahedron.

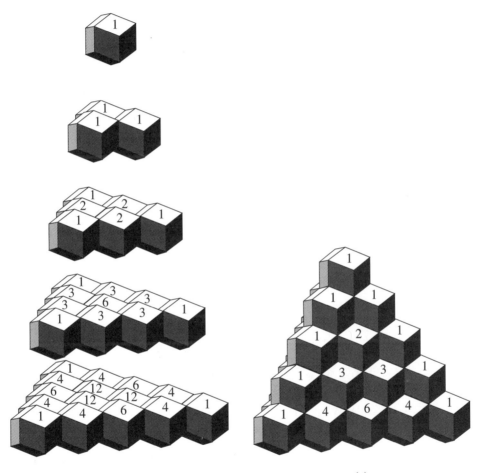

(b)   (c)

FIGURE 14 (*continued*)

Symmetry Identity

$$\begin{pmatrix} n \\ r\ s\ t \end{pmatrix} = \begin{pmatrix} n \\ r\ t\ s \end{pmatrix} = \begin{pmatrix} n \\ s\ r\ t \end{pmatrix} = \begin{pmatrix} n \\ s\ t\ r \end{pmatrix} = \begin{pmatrix} n \\ t\ r\ s \end{pmatrix} = \begin{pmatrix} n \\ t\ s\ r \end{pmatrix}$$

(24)

We also have the

Pascal Identity

$$\begin{pmatrix} n \\ r\ s\ t \end{pmatrix} = \begin{pmatrix} n-1 \\ r-1\ s\ t \end{pmatrix} + \begin{pmatrix} n-1 \\ r\ s-1\ t \end{pmatrix} + \begin{pmatrix} n-1 \\ r\ s\ t-1 \end{pmatrix}$$

(25)

where $n, r, s, t \geq 1$.

Figure 14(b) shows layers of polyhedra, well-known as ***regular rhombic dodecahedra***, stacking up to form the Pascal Tetrahedron, just as Figure 1(a) shows regular hexagons stacking up to form the Pascal Triangle.[13] Hence the Pascal Identity may be expressed by saying that the sum of the three numbers identified with the rhombic dodecahedra touching the "roof" of a rhombic dodecahedron equals the number assigned to that rhombic dodecahedron. (Remember our remark relating to the Pascal Identity for binomial coefficients.)

• • • **BREAK 3**

(1) Prove (25) by means of a combinatorial argument. (Hint: Think of putting n students into three classrooms so that the first, second, and third classrooms contain r, s, t students, respectively. Then do the same thing, but with exactly one student wearing a hat!)

(2) Give either an arithmetic proof (this can be quite neat!) or an algebraic proof of (25).

(3) What do you notice about the numbers on the slanted faces of the Pascal Tetrahedron?

When we display the numerical values of $\begin{pmatrix} n \\ r\ s\ t \end{pmatrix}$ it is most natural to place them in a tetrahedral array, with triangular cross sections, as shown in Figure 14(a). You will then notice that there are planes in each of which one of the four variables n, r, s, t is constant. Of course, for an expansion of $(a + b + c)^n$ in a particular case, we would actually prefer to use the numerical values that appear at the appropriate level of an extended version of Figure 14(a). The clever layout inside the bubble of Figure 14(a) was

---

[13] A particularly satisfying part of our analogy is the fact that regular hexagons fill the plane and regular rhombic dodecahedra fill space.

discovered and shown to us by Larry Pierce II, at that time a sophomore student at the University of Toledo. To see *why* it is true, consider a particular but not special case, say $(a + b + c)^3$. Then observe that, using the binomial expansion, we have

$$(a + (b + c))^3 = a^3 + 3a^2(b + c) + 3a(b + c)^2 + (b + c)^3.$$

Compare this with the diagram inside the bubble for level 3 in Figure 14(a). The models shown in Figures 14(b) and (c) may be constructed using rhombic dodecahedra. A 3-dimensional model and the illustrations in Figures 14(b) and (c) were created for us by our colleague Hans Walser. He is currently preparing an article on the Pascal Tetrahedron, which involves, among other things, some of the geometric properties of the rhombic dodecahedron, and describes how to build this particular model of the Pascal Tetrahedron (which he calls the Pascal Pyramid) using a collection of rhombic dodecahedra.

● ● ● **BREAK 4**

(1) Use the values at level $n = 4$ in the Pascal Tetrahedron to fill in the coefficients for the expansion of $(2x - y + 3z)^4$. (Hint: It may be helpful to write out the following array:

$$\underline{\quad}(2x)^4$$
$$+ \underline{\quad}(2x)^3(-y) + \underline{\quad}(2x)^3(3z)$$
$$+ \underline{\quad}(2x)^2(-y)^2 + \underline{\quad}(2x)^2(-y)(3z) + \underline{\quad}(2x)^2(3z)^2$$
$$+ \underline{\quad}(2x)(-y)^3 + \underline{\quad}(2x)(-y)^2(3z) + \underline{\quad}(2x)(-y)(3z)^2 + \underline{\quad}(2x)(3z)^3$$
$$+ \underline{\quad}(-y)^4 + \underline{\quad}(-y)^3(3z) + \underline{\quad}(-y)^2(3z)^2 + \underline{\quad}(-y)(3z)^3 + \underline{\quad}(3z)^4$$

(2) What is the coefficient of $a^3b^4c^2d^5$ in the expansion of $(a + 2b - 3c + d)^{14}$?

Figure 14 suggests that there may be numerical patterns that are linked with the geometry of this array of numbers that are analogous to our Generalized Star of David Theorems for binomial coefficients. Since we are certain that many of our readers will want to look at the generalization to **multinomial (m-nomial) coefficients**, $m \geq 3$, we will use a notation for the trinomial coefficients that sets the scene for such a generalization.

Thus, anticipating the more general case, we will write $\binom{n}{k_1 \ k_2 \ k_3}$ instead of the more usual $\binom{n}{r \ s \ t}$. Now, how do we know *which* geometric configuration will be of interest to us in finding analogues to our results for binomial coefficients? Actually we don't know for sure, but since the genesis of our Star of David theorems was a parallelogram whose sides were drawn in certain specific directions, we hope it will be fruitful to consider a fundamental parallelepiped[14] $P^{(n)}$, with vertices

$$
\begin{cases}
A\begin{pmatrix} n \\ k_1 \ k_2 \ k_3 \end{pmatrix}, & C\begin{pmatrix} n + a_1 + a_2 \\ k_1 + a_1 \quad k_2 + a_2 \quad k_3 \end{pmatrix}, \\[2mm]
E\begin{pmatrix} n + a_2 + a_3 \\ k_1 \quad k_2 + a_2 \quad k_3 + a_3 \end{pmatrix}, & G\begin{pmatrix} n + a_1 + a_3 \\ k_1 + a_1 \quad k_2 \quad k_3 + a_3 \end{pmatrix} \\[2mm]
\text{- -} \\[2mm]
B\begin{pmatrix} n + a_1 \\ k_1 + a_1 \quad k_2 \quad k_3 \end{pmatrix}, & D\begin{pmatrix} n + a_2 \\ k_1 \quad k_2 + a_2 \quad k_3 \end{pmatrix}, \\[2mm]
F\begin{pmatrix} n + a_3 \\ k_1 \quad k_2 \quad k_3 + a_3 \end{pmatrix}, & H\begin{pmatrix} n + a_1 + a_2 + a_3 \\ k_1 + a_1 \quad k_2 + a_2 \quad k_3 + a_3 \end{pmatrix}
\end{cases}
\tag{26}
$$

as shown in Figure 15 (where we assume that each of the three axes points in the positive direction). We refer to a_1, a_2, a_3 as the *lengths* of the edges of the parallelepiped. Notice that each of the faces of this parallelepiped lies in a plane for which one of k_1, k_2, k_3 is a constant.

We describe the vertices A, C, E, G as the *even* vertices, and B, D, F, H as the *odd* vertices, because the number of different edges traversed between the anchor A and the vertex in question (regardless of the length of the edge) is an even number for A, C, E, G and an odd number for B, D, F, H. This fact may be seen by counting the number of edges between the anchor and the vertex in question in Figure 15 (or the Walser model 15A). Equivalently, the parity may be determined by counting the number of k_i's that are changed (by the addition of a suitable a_i) in the bottom of the symbol $\begin{pmatrix} n \\ k_1 \ k_2 \ k_3 \end{pmatrix}$ to achieve the multinomial coefficient in question. This method is consistent with what we have already done with binomial coefficients; and will certainly be applicable to the general m-nomial $\begin{pmatrix} n \\ k_1 \ k_2 \ \cdots \ k_m \end{pmatrix}$.

Returning to trinomial coefficients, the weight of $P^{(n)}$ is given by

$$
W(P^{(n)}) = \frac{\text{product of even vertices}}{\text{product of odd vertices}} = \frac{A \times C \times E \times G}{B \times D \times F \times H}
$$

so that, by a straightforward calculation,

$$
W(P^{(n)}) = \frac{n!(n + a_1 + a_2)!(n + a_2 + a_3)!(n + a_1 + a_3)!}{(n + a_1)!(n + a_2)!(n + a_3)!(n + a_1 + a_2 + a_3)!}
\tag{27}
$$

[14]A parallelepiped is just a squashed rectangular box.

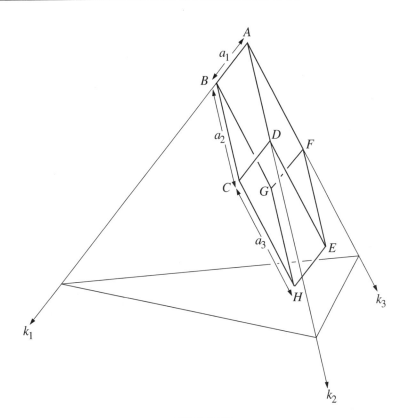

FIGURE 15A

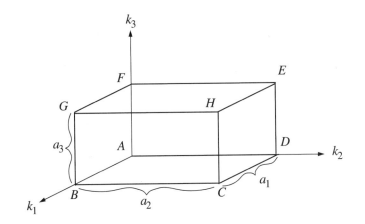

FIGURE 15 The Fundamental Parallelepiped $P^{(n)}$.

(Compare the form of (27) with the form of the weight in the binomial case, given in the footnote accompanying BREAK 2.) Thus, for fixed a_1, a_2, a_3, $W(P^{(n)})$ is independent of k_1, k_2, k_3 and is therefore invariant under the action of sliding $P^{(n)}$ in the plane $n = $ constant. Further, we can permute the triple (a_1, a_2, a_3) in any way we wish and the weight will remain unchanged; that is, $W(P^{(n)})$ is invariant under the action of the symmetric group S_3 on the set of coordinate directions (or, equivalently, on the set $\{a_1, a_2, a_3\}$). Thus, with $m = 3$, we have $5(= 3! - 1)$ other parallelepipeds having the same weight as $P^{(n)}$.

Notice that we could have taken some of a_1, a_2, a_3 to be negative, but we would have had to ensure that, in every factorial on the right of (27), we were dealing with non-negative numbers. In particular, if all a_1, a_2, a_3 are replaced by their negatives, we get

$$W(P^{(n)}) = \frac{n!(n - a_1 - a_2)!(n - a_2 - a_3)!(n - a_1 - a_3)!}{(n - a_1)!(n - a_2)!(n - a_3)!(n - a_1 - a_2 - a_3)!} \quad (28)$$

which would then look completely analogous to the answer for $W(P^{(n)})$ in the binomial case (see BREAK 2).

Figure 16 illustrates the case where we consider the cyclic permutation π in which a_1, a_2, a_3 go to a_3, a_1, a_2, respectively. Let $\overset{*}{P}{}^{(n)}$ be the π-image[15] of $P^{(n)}$, with its vertices labeled in the obvious way (as shown in Figure 16). Then A is invariant under the permutation π, and we get a **Hyperstar of David Theorem** of the form

$$B^* \times C \times D^* \times E \times F^* \times G \times H^* = B \times C^* \times D \times E^* \times F \times G^* \times H \quad (29)$$

Since π is a cyclic permutation of length 3 (as illustrated in Figure 16 and the Walser model Figure 16A), then (29) is a genuinely new identity.

However, if we choose a permutation π that leaves some a_j fixed, then (29) degenerates to a disguised form of (20). Thus if π simply exchanges a_1 and a_2, then π leaves both the anchor A and the vertex F fixed ($F^* = F$); and then (29) becomes a simple consequence of the two binomial Star of David identities

$$B \times C^* \times D = B^* \times C \times D^* \quad \text{and} \quad E \times G \times H^* = E^* \times G^* \times H \quad (30)$$

at levels k_3 and $k_3 + a_3$, respectively (see Figure 17).

What we have said about the parallelepiped $P^{(n)}$ may be modified to deal with parallelepipeds $P_{(k_1)}, P_{(k_2)}, P_{(k_3)}$. The next break will give you the opportunity to see how this is carried out.

---

[15]Thus $\overset{*}{P}{}^{(n)}$ is one of the 5 other parallelepipeds referred to above.

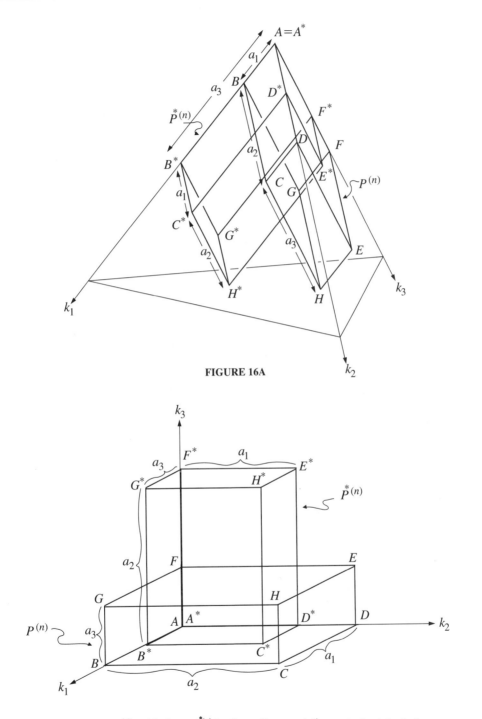

FIGURE 16A

FIGURE 16 $P^{(n)}$ and its image $\overset{*}{P}{}^{(n)}$ for the cyclic permutation $a_1, a_2, a_3 \rightarrow a_3, a_1, a_2$.

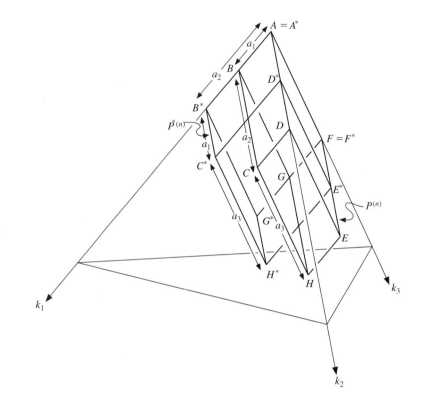

FIGURE 17A

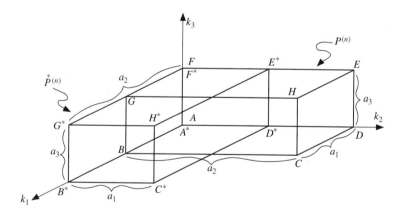

FIGURE 17 $P^{(n)}$ and its image $\overset{*}{P}{}^{(n)}$ for the permutation that exchanges a_1 and a_2.

• • • **BREAK 5**

(1) Let $\left(\begin{smallmatrix} & n & \\ k_1 & k_2 & k_3 \end{smallmatrix}\right)$ be the anchor of a parallelepiped with edge lengths a_1, a_2, a_3 in the hyperplanes where n, k_2, k_3, respectively, are constant. Show that

$$W(P_{(k_1)}) = \frac{(k_1 - a_1)!(k_1 - a_2)!(k_1 - a_3)!(k_1 - a_1 - a_2 - a_3)!}{k_1!(k_1 - a_1 - a_2)!(k_1 - a_1 - a_3)!(k_1 - a_2 - a_3)!}$$

(Hint: Use Figure 15, but replace the k_1-axis with n, running in the negative direction. Leave the k_2- and k_3-axes running in the positive direction. Thus,

$$A = \begin{pmatrix} & n & \\ k_1 & k_2 & k_3 \end{pmatrix},$$

$$B = \begin{pmatrix} & n - a_1 & \\ k_1 - a_1 & k_2 & k_3 \end{pmatrix},$$

$$C = \begin{pmatrix} & n - a_1 & \\ k_1 - a_1 - a_2 & k_2 + a_2 & k_3 \end{pmatrix},$$

$$\vdots$$

(2) Find $W(P_{(k_2)})$ and $W(P_{(k_3)})$.

(3) Compare the forms for these three weights with each other and with (27) or (28).

(4) Guess what $W(P^{(n)})$ and $W(P_{(k_1)})$ would be for tetranomial coefficients ($m = 4$).

Of course, we can generalize the material of this chapter to m-multinomial[16] coefficients. That would be a story well worth pursuing, but, instead, we will take up the problem of extending the trinomial coefficients beyond the Pascal Tetrahedron.

---

[16] Already in the case $m = 4$ two additional features, very different in nature, make their appearance. First, there are permutations π in S_4, for example $a_1, a_2, a_3, a_4 \to a_2, a_1, a_4, a_3$, that are neither cyclic nor such that some a_i is fixed. Second, we can no longer draw pictures to interrelate the geometry and the algebra. Thus, as the situation to be explored gets richer, our available tools get poorer!

6.3   EXTENDING THE PASCAL TETRAHEDRON
AND THE PASCAL m-SIMPLEX

We will argue in this section that the rule for defining trinomial coefficients

$$\begin{pmatrix} & n & \\ r & s & t \end{pmatrix},$$

where n is a *negative* integer, should be as follows:

$$\begin{cases} \begin{pmatrix} & n & \\ r & s & t \end{pmatrix} = 0, \quad \text{unless} \\[2ex] r \geq 0, \; s \geq 0, \quad \text{when } \begin{pmatrix} & n & \\ r & s & t \end{pmatrix} = (-1)^{n-t} \begin{pmatrix} & -t-1 & \\ r & s & -n-1 \end{pmatrix}, \\[2ex] \text{or} \;\; r \geq 0, \; t \geq 0, \quad \text{when } \begin{pmatrix} & n & \\ r & s & t \end{pmatrix} = (-1)^{n-s} \begin{pmatrix} & -s-1 & \\ r & -n-1 & t \end{pmatrix}, \\[2ex] \text{or} \;\; s \geq 0, \; t \geq 0, \quad \text{when } \begin{pmatrix} & n & \\ r & s & t \end{pmatrix} = (-1)^{n-r} \begin{pmatrix} & -r-1 & \\ -n-1 & s & t \end{pmatrix}. \end{cases}$$

$$(31)$$

We base this claim on the observation that, if $(a + b + c)^n$, with n *negative*, is to be represented by a convergent power series, then there are only 3 possibilities (compare [4, 5] and rule (10) above). We will have

$$(a+b+c)^n = \sum_{r,s \geq 0} \begin{pmatrix} & n & \\ r & s & t \end{pmatrix} a^r b^s c^t \qquad (32)$$

or

$$(a+b+c)^n = \sum_{r,t \geq 0} \begin{pmatrix} & n & \\ r & s & t \end{pmatrix} a^r b^s c^t \qquad (33)$$

or

$$(a+b+c)^n = \sum_{s,t \geq 0} \begin{pmatrix} & n & \\ r & s & t \end{pmatrix} a^r b^s c^t \qquad (34)$$

where the coefficients $\begin{pmatrix} n \\ r\, s\, t \end{pmatrix}$ are given by the appropriate formulae in (31). We will be content to show that the formula (32) yields a convergent power series in the open subset of $\mathbb{R}^3$ given by $|a + b| < c,\ |a| < |c|,\ |b| < |c|$.

For, if $|a + b| < c$, then, by the rule given in [4, 5, 6, 7], which coincides with rule (10), we have

$$(a + b + c)^n = \sum_{\substack{u \geq 0 \\ u+t=n}} \binom{n}{u \ \ t} (a + b)^u c^t,$$

where

$$\binom{n}{u \ \ t} = (-1)^u \binom{-t-1}{u \ \ -n-1}.$$

Now we may expand $(a + b)^u$ by the binomial theorem and obtain a convergent double power series

$$(a + b + c)^n = \sum_{\substack{r,s \geq 0 \\ r+s=u \\ u+t=n}} \binom{u}{r \ \ s}\binom{n}{u \ \ t} a^r b^s c^t$$

and

$$\binom{u}{r \ \ s}\binom{n}{u \ \ t} = (-1)^u \binom{u}{r \ \ s}\binom{-t-1}{u \ \ -n-1}$$

$$= (-1)^{n-t} \binom{-t-1}{r \ \ s \ \ -n-1}$$

Thus we obtain formula (32).

★ Obviously, the rules (31) through (34) extend to m-multinomial coefficients. We start, of course, with the m-dimensional analogue of the triangle and tetrahedron, namely, the m-dimensional simplex, or m-simplex (see [8]). In the Pascal m-simplex, with $n, k_1, k_2, \cdots, k_m$ non-negative and $\sum_i k_i = n$, we have, as expected,

$$\binom{n}{k_1 \ \ k_2 \ \ \cdots \ \ k_m} = \frac{n!}{k_1! k_2! \cdots k_m!} \tag{35}$$

Now let n be a negative integer. Then, without going into details, we obtain the following rule: With n a negative integer,

$$\binom{n}{k_1 \ \ k_2 \ \ \cdots \ \ k_m} = 0,$$

unless *exactly* one of $k_1, k_2, \cdots, k_m$ is negative. If k_i is negative, then

$$
\binom{n}{k_1 \ k_2 \ \cdots \ k_m} = (-1)^{n-k_i} \binom{-k_i - 1}{k_1 \ \cdots \ k_{i-1} \ (-n-1) \ k_{i+1} \ \cdots \ k_m}
$$

(36)

It is interesting to remark that (36) generalizes the rule for defining $\binom{n}{r \ s}$ in the part of the Pascal Hexagon given by $n < 0$, $r \geq 0$, $s < 0$, (see Figure 4), namely,

$$
\binom{n}{r \ s} = (-1)^r \binom{-s-1}{r \ \ -n-1}
$$

However, to construct the generalization, notice that we must replace $(-1)^r$ above by the equivalent $(-1)^{n-s}$. Of course, (36) also generalizes (31) and is used in the same way — to calculate the LHS by means of the known arithmetic value of the RHS (see (35)).

6.4   SOME VARIANTS AND GENERALIZATIONS

In [9] Hoggatt and Alexanderson give an interesting algorithm relating to the multinomial coefficients. We again take an anchor $\binom{n}{k_1 \ k_2 \ \cdots \ k_m}$ and consider its *nearest neighbors*. Precisely, we have, at level $(n + 1)$, the m multinomial coefficients

$$
\binom{n+1}{k_1 + 1 \ k_2 \ \cdots \ k_m}, \qquad \binom{n+1}{k_1 \ k_2 + 1 \ \cdots \ k_m},
$$

$$
\cdots, \qquad \binom{n+1}{k_1 \ k_2 \ \cdots \ k_m + 1}
$$

at level n we have the $m(m-1)$ multinomial coefficients

$$
\binom{n}{k_1 \ k_2 \ \cdots \ k_{i-1} \ k_i + \epsilon \ \cdots \ k_j - \epsilon \ \cdots \ k_m},
$$

$$
1 \leq i < j \leq m, \quad \epsilon = \pm 1
$$

and at level $(n-1)$ we have the m multinomial coefficients

$$
\binom{n-1}{k_1-1 \quad k_2 \quad \cdots \quad k_m}, \qquad \binom{n-1}{k_1 \quad k_2-1 \quad \cdots \quad k_m},
$$

$$
\cdots, \qquad \binom{n-1}{k_1 \quad k_2 \quad \cdots \quad k_m-1}
$$

These $m(m+1)$ coefficients, as Hoggatt and Alexanderson prove, may then be partitioned into m sets, each containing $(m+1)$ coefficients, in such a way that the product of the coefficients in each set is a given number N, namely[17]

$$
N = \frac{(n-1)!(n!)^{m-1}(n+1)!}{\prod(k_i-1)!\left(\prod k_i!\right)^{m-1}\prod(k_i+1)!} \tag{37}
$$

which is thereby seen to be an integer. The method of partitioning is not claimed to be unique. This theorem may, in the case $m=2$, be regarded as a restatement of the Star of David Theorem (20), in the special case $k=1$, $\ell=-1$, which is the original case of the Star of David Theorem discussed by Gould [2]; notice, again, that we may certainly allow negative values for k and ℓ in the Star of David Theorem, so long as all entries in the binomial coefficients in (20) remain non-negative. However, for $m \geq 3$, the Hoggatt–Alexanderson Theorem [9] diverges from what we have called the Hyperstar of David Theorem.

On the other hand, we may generalize the Hoggatt–Alexanderson Theorem in the following direction. We regard the theorem we have described as the case $(1, -1)$ of the following theorem.

Theorem 1 *Given the multinomial anchor $\binom{n}{k_1 \; k_2 \; \cdots \; k_m}$, we define its set of (p,q)-satellites[18] to consist at level $(n+p)$ of the m coefficients*

$$
\binom{n+p}{k_1+p \quad k_2 \quad \cdots \quad k_m}, \qquad \binom{n+p}{k_1 \quad k_2+p \quad \cdots \quad k_m},
$$

$$
\cdots, \qquad \binom{n+p}{k_1 \quad k_2 \quad \cdots \quad k_m+p}
$$

---

[17]Here, and in (38), $\prod r_i!$ is an abbreviation for $\prod(r_i!)$, and, of course, $\prod$ means the product over all values of i.

[18]Thus, what we called *nearest neighbors* are the $(1, -1)$-satellites.

at level $(n + q)$ of the m coefficients

$$\binom{n+q}{k_1+q \quad k_2 \ \cdots \ k_m}, \qquad \binom{n+q}{k_1 \quad k_2+q \ \cdots \ k_m},$$

$$\cdots, \qquad \binom{n+q}{k_1 \quad k_2 \ \cdots \ k_m+q}$$

and at level $(n + p + q)$ of the $m(m - 1)$ coefficients

$$\binom{n+p+q}{k_1 \quad k_2 \ \cdots \ k_{i-1} \ k_i+\epsilon \ k_{i+1} \ \cdots \ k_{j-1} \ k_j+\eta \ k_{j+1} \ \cdots \ k_m},$$

$$1 \le i < j \le m, \quad (\epsilon, \eta) = (p, q) \text{ or } (q, p).$$

Then these $m(m + 1)$ coefficients may be partitioned into m sets, each containing $(m + 1)$ coefficients, in such a way that the product of the coefficients in each set is the number $N(p, q)$ given by

$$N(p, q) = \frac{(n+p)!(n+q)!\,((n+p+q)!)^{m-1}}{\prod(k_i+p)! \prod(k_i+q)! \left(\prod k_i!\right)^{m-1}} \tag{38}$$

Corollary 2   *The number $N(p, q)$ is an integer.*

Proof $N(p, q)$ is a rational number whose mth power is an integer.   □

Note that p, q may take any integer values such that the factorial functions in (38) are applied to non-negative integers.

Of course, the partitioning function is exactly that used by the authors of [9]; theirs is the hard work!

We now discuss an entirely different variant of our earlier results. In this variant we generalize the content of Section 2 substantially beyond the domain of multinomial coefficients. Indeed, all we required in that section of the multinomial coefficient, as a function of $n, k_1, k_2, \cdots, k_m$, is that it be a *separable* function of its variables; that is, we may replace the multinomial coefficient by any such function:

$$F(n, k_1, k_2, \cdots, k_m) = f(n) f_1(k_1) \cdots f_m(k_m) \tag{39}$$

An example of such a separable function is the **q-analogue** of the multinomial coefficient. This is the Gaussian polynomial (see [10]), obtainable from the multinomial coefficient $\binom{n}{k_1 \ k_2 \ \cdots \ k_m}$ by replacing each occurrence of $r!$ by $\frac{(q^r-1)(q^{r-1}-1)\cdots(q-1)}{(q-1)^r}$; of course, $0!_q$ is just 1. The Gaussian polynomial

$\left(\begin{smallmatrix} & & n & \\ k_1 & k_2 & \cdots & k_m \end{smallmatrix}\right)_q$ may be regarded as the coefficient of $a_1^{k_1} a_2^{k_2} \cdots a_m^{k_m}$ in the expansion of $(a_1 + a_2 + \cdots + a_m)^n$ in the algebra (over $\mathbb{R}$) generated by $(a_1, a_2, \cdots, a_m; q)$, subject to $a_j a_i = q a_i a_j$, $i < j$, $q a_i = a_i q$. Notice that, in this generalization of the material of Section 2, it is not necessary that the functions $f_1, f_2, \cdots, f_m$ be the same; however, if F in (39) is (as in the special case of the multinomial coefficients and in the example of the Gaussian polynomials) invariant under the action of the symmetric group S_m on the m-tuple $(k_1, k_2, \cdots, k_m)$, then the functions $f_1, f_2, \cdots, f_m$ in (39) may be chosen to be the same.

The material of Section 3 makes it obvious how we would extend the definition of F, given by (39), to the domain of arbitrary integers $n, k_1, k_2, \cdots, k_m$. The formula (35) is easily generalized by replacing the multinomial coefficient by F; indeed, the generalization does not even require that F be separable. However, one has to be careful in assigning a value to F in the regions in which the multinomial coefficient takes the value 0; this problem was first encountered in extending the harmonic triangle in [4, 7]. We do not want to put any stress on this technical point here, since it is in the regions where formula (35) holds that the real interest lies.

Finally, we may obviously generalize Theorem 1, in Section 4, provided that we insist in (39) that F be invariant under the action of S_m. We leave the details to the reader.

Another direction to take, starting from the Star of David Theorem is ★ suggested in [11].

6.5 THE GEOMETRY OF THE 3-DIMENSIONAL ANALOGUE OF THE PASCAL HEXAGON

We revert here to the notation $\left(\begin{smallmatrix} & n & \\ r & s & t \end{smallmatrix}\right)$ for the trinomial coefficient, so that we can compare the analogue more easily with the Pascal Hexagon[19] in Figure 1. To see how we might find the appropriate polyhedron to accommodate the trinomial coefficients, when n, r, s, t are any integers subject to $r + s + t = n$, we make the following observations about the Pascal Hexagon in two dimensions (see Figures 18 and 19, which are reproductions of Figures 3 and 4, respectively, with some subtle embellishments):

1. There are precisely $2^3 - 2 = 6$ regions in the plane, since it is possible to have all combinations of assignments of signs to n, r, s except $n \geq 0$, $r < 0, s < 0$ and $n < 0, r \geq 0, s \geq 0$.

---

[19]It is also the case that we do not know the m-dimensional analogue of the Pascal Hexagon for $m > 3$.

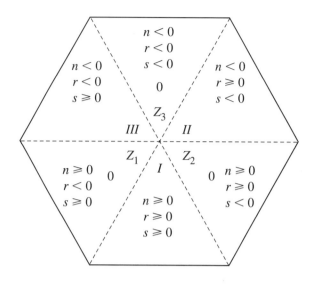

FIGURE 18   Notice that precisely one inequality changes whenever a line between sextants of the hexagon is crossed.

2. The lines separating the six sextants may be thought of as the extensions of the sides of an equilateral triangle having 1's at each of its vertices, as shown in Figure 19.

3. The three non-zero sextants are arranged as symmetrically as possible, and each one contains two boundaries emanating from the center.

4. Precisely one of the three inequalities changes whenever a line separating sextants of the hexagon is crossed (see Figure 18).

So we begin our search for the 3-dimensional analogue by observing that, for the trinomial coefficients $\left(\begin{smallmatrix} & n & \\ r & s & t \end{smallmatrix}\right)$, we have, by analogy with observation (1) above, the property: There are precisely $2^4 - 2 = 14$ regions in space, since it is possible to have all combinations of assignments of sign to n, r, s, t except $n \geq 0, r < 0, s < 0, t < 0$ and $n < 0, r \geq 0, s \geq 0, t \geq 0$.

Now, what we need is a symmetric polyhedron whose extended face planes produce 14 distinct regions in space satisfying the analogue of observation (2) above in a satisfactory way. So we ask first, what is the 3-dimensional analogue of an equilateral triangle? The natural answer is the regular tetrahedron (and we know that, when $n, r, s, t \geq 0$, the trinomial coefficients can be arranged in a regular tetrahedral region as illustrated in

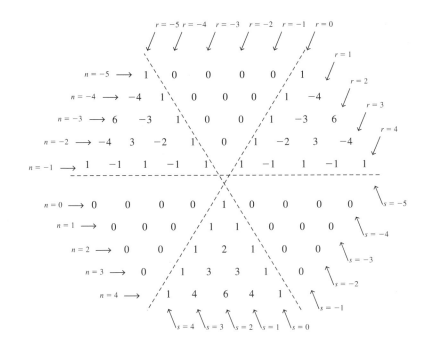

FIGURE 19 The Pascal Hexagon, showing values of $\binom{n}{r\ s}$, with $r + s = n$, for $-5 \le n, r, s \le 4$.

Figure 1). Now if we extend the face planes of a regular tetrahedron we obtain

1. 4 trihedral regions off the vertices,
2. 4 truncated trihedral regions off the faces, and
3. 6 wedges off the edges (with rectangular cross sections, tending to square cross sections in the limit).

Voilà! It turns out that, when a centrally symmetric boundary is placed around the configuration with flat faces covering each region, the extended face planes of the regular tetrahedron tend[20] to produce a polyhedron with eight equilateral triangular faces and six square faces. This is, in fact, the cuboctahedron. Can the cuboctahedron satisfy the rest of our conditions? We begin our answer by conjecturing that the four non-zero regions are trihedral regions off the vertices of a regular tetrahedron with 1's at each vertex and the other regions are the zero regions. We check to see if this is so

---

[20]Here, just as in the original case of the hexagon, one moves towards perfect regularity as the values of n, r, s, t tend to $\pm\infty$.

and to see if we get an arrangement of faces (representing regions) that gives situations consistent with the analogous statements of (3) and (4). The net diagram of the cuboctahedron in Figure 20 shows that it is, indeed, possible to arrange the 14 sets of inequalities so that precisely one inequality changes whenever an edge of the polyhedron is crossed.

By copying the diagram in Figure 20 and actually constructing the model (it is advisable to draw a tab on every other edge on the exterior boundary

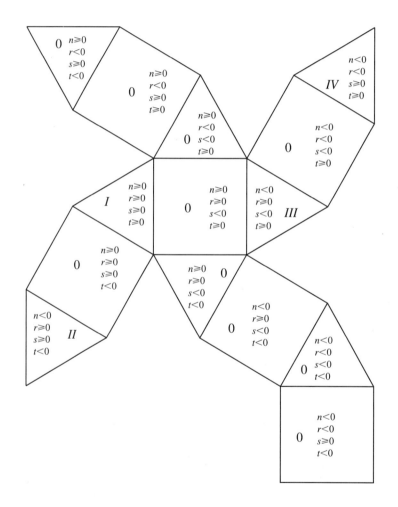

FIGURE 20   Net diagram for the Pascal Cuboctahedron. Notice that precisely one inequality changes whenever an edge of the polyhedron is crossed.

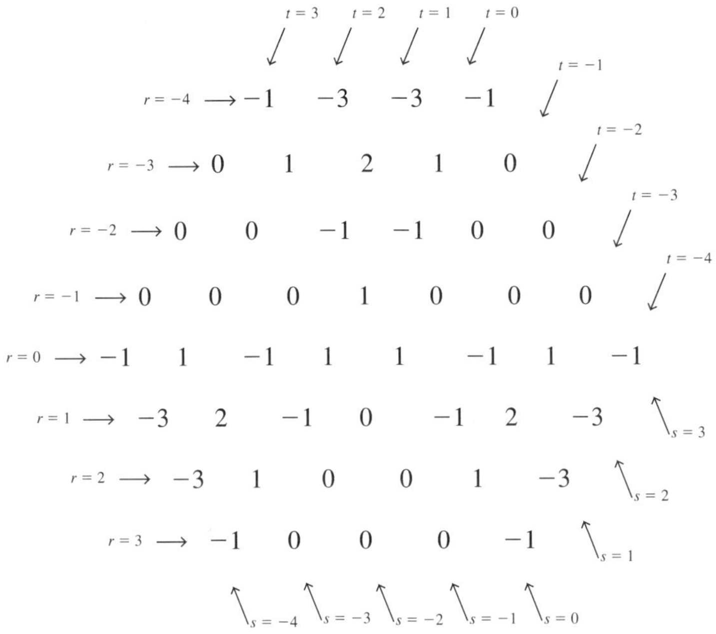

FIGURE 21 Hexagonal cross-section of the Pascal Cuboctahedron when $n = -1$ and $-4 \leq r, s, t \leq 3$.

before cutting the pattern out), one sees that the non-zero regions are located so that they may be thought of as occupying the four trihedral regions off the vertices of the central tetrahedron. But there is even more confirmation that this is the right polyhedron. By examining the four hexagonal cross-sections of this model one sees that when $t = 0$ the cross-section actually is the Pascal Hexagon! Moreover, the two other hexagons, obtained when $s = 0$ or when $r = 0$, have the identical numerical values. A fourth non-zero hexagonal cross-section occurs when $n = -1$, and some of its values are shown in Figure 21.

This is certainly the polyhedron we seek. We call it the **Pascal Cuboctahedron**.[21] We rest our case!

---

[21] A clever pop-up model of the Pascal Cuboctahedron was very kindly and very ingeniously invented by Hans Walser, at the request of PH and JP, to enable them to lecture on this topic when away from their home bases. In addition to the article [13] about the Pascal Tetrahedron (which he calls the Pascal Pyramid), referred to in Section 2, he has also written an article [12] with instructions on how to build this pop-up model.

REFERENCES

1. Aldred, Robert, Peter Hilton, Derek Holton, and Jean Pedersen, Baked Beans and Spaghetti, *The Mathematics Educator* **5**, No. 2 (1994), 35–41.

2. Gould, H.W., Equal products of generalized binomial coefficients, *Fibonacci Quarterly* **9** (1971), 337–346.

3. Hilton, Peter, Derek Holton, and Jean Pedersen, *Mathematical Reflections — In a Room with Many Mirrors*, 2nd printing, Springer Verlag NY, 1998.

4. Hilton, Peter, and Jean Pedersen, Extending the binomial coefficients to preserve symmetry and pattern, *Computers Math. Applic.* **17**:1–3 (1989) 89–102.

5. Hilton, Peter, and Jean Pedersen, Relating geometry and algebra in the Pascal triangle, hexagon, tetrahedron, and cuboctahedron. Part I: Binomial coefficients, extended binomial coefficients and preparation for further work, *The College Mathematics Journal*, **30**, No. 3 (1999), 170–186.

6. Hilton, Peter, and Jean Pedersen, Relating geometry and algebra in the Pascal triangle, hexagon, tetrahedron, and cuboctahedron. Part II: Geometry and algebra in higher dimensions: Identifying the Pascal cuboctahedron, *The College Mathematics Journal*, **30**, No. 4 (1999), 279–292.

7. Hilton, Peter, Jean Pedersen, and William Rosenthal, Pascalian triangles and extensions to hexagons, *Quaestiones Mathematicae* **13**, No. 3-4 (1990), 395–416.

8. Hilton, Peter, and Shaun Wylie, *Homology Theory*, Cambridge University Press, 3rd printing, 1965.

9. Hoggatt, V.E. Jr, and G.L. Alexanderson, A property of multinomial coefficients, *Fibonacci Quarterly* **9** (1971), 351–356, 420.

10. Pólya, G., and G.L. Alexanderson, Gaussian binomial coefficients, *Elemente der Mathematik* **26**, No. 5 (1991), 102–108.

11. Usiskin, Zalman, Perfect square patterns in the Pascal triangle, *Mathematics Magazine*, **46**, No. 4 (1973), 203–208.

12. Walser, Hans, The Pop-up Cuboctahedron, *The College Mathematics Journal*, **31**, No. 2 (2000), 89–92.

13. Walser, Hans, The Pascal Pyramid, *The College Mathematics Journal*, **31**, No. 5 (2000), 383–392.

7 | Catalan Numbers

C H A P T E R

7.1 INTRODUCTION: THREE IDEAS
ABOUT THE SAME MATHEMATICS

The *classical* Catalan numbers, rediscovered by the Belgian mathematician Eugène Charles Catalan (1814–1894) seem to have been first studied by the famous Swiss mathematician Leonhard Euler (1707–1783). Euler considered the problem of counting the number of ways a given convex polygon[1] can be divided into triangles by drawing non-intersecting diagonals (a diagonal is a line segment joining non-adjacent vertices, and two diagonals are considered to be non-intersecting if they intersect only in a vertex of the polygon). Obviously, this number depends only on n, the number of sides of the polygon; obviously, too, $(n-3)$ diagonals will be drawn, creating $(n-2)$ triangles. We will call the number[2] c_{n-2}, thus putting emphasis on the number of triangles; this accords with standard practice. It is easy to see that, for polygons with $3, 4, 5$ sides, we have $c_1 = 1$, $c_2 = 2$, $c_3 = 5$, respectively (see Figure 1). Even Euler, however, found it difficult to obtain a general formula for c_k.

---

[1] Although we will always draw our polygons as regular polygons, this is not necessary. All that is necessary is that the polygons be convex. We also assume that the given polygon cannot be rotated or reflected — otherwise all the polygons, for a given value of n, in Figure 1 would be equivalent to each other.

[2] We will write c_k for c_{n-2}, and, conventionally, set $c_0 = 1$.

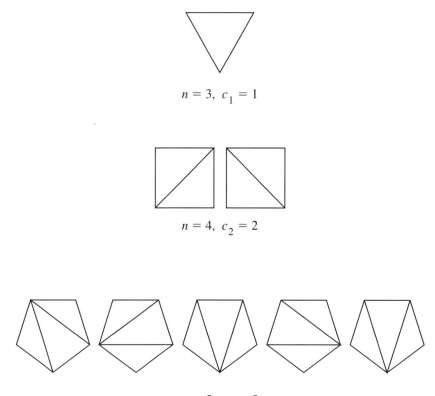

$n = 3, \; c_1 = 1$

$n = 4, \; c_2 = 2$

$n = 5, \; c_3 = 5$

FIGURE 1

● ● ●   **BREAK 1**

Draw some hexagons and divide them into triangles by non-intersecting diagonals to determine the value of c_4.

Today, we know many different interpretations of the classical Catalan numbers. Henry Gould, a mathematician at the University of West Virginia, has available a huge bibliography containing 450 references to Catalan numbers (see [2]) and involving a very large number of interpretations of the sequence of (classical) Catalan numbers. See also [8, 9].

We plan in this chapter to give you four interpretations (each preceded by a bullet (●), so that they stand out on the page). You have one already; let us repeat it in more convenient form:

- $c_0 = 1$; c_k, $k \geq 1$, is defined to be the number of ways of partitioning a $(k + 2)$-gon into k triangles by means of $(k - 1)$ non-intersecting diagonals.

Our second interpretation of the classical Catalan numbers is concerned with a given binary operation on a set — we will simply write the operation as juxtaposition. Then

- $b_0 = 1$; b_k, $k \geq 1$, is defined to be the number of ways of introducing parentheses into (or, as we may say, **parenthesizing**) the symbol $s_1 s_2 \cdots s_{k+1}$ so that it makes sense.

Plainly, $b_1 = 1$, since the only possibility is $(s_1 s_2)$. Then $b_2 = 2$, since we may have $((s_1 s_2) s_3)$ or $(s_1 (s_2 s_3))$.

• • • **BREAK 2**

(1) Write out all the ways of parenthesizing the symbol $s_1 s_2 s_3 s_4$ to determine the value of b_3.
(2) Write out all the ways of parenthesizing the symbol $s_1 s_2 s_3 s_4 s_5$ to determine the value of b_4.
(3) Do you see a systematic way of going from b_{k-1} to b_k?

Your answers to BREAKS 1 and 2 should convince you of two things: first, that it is a reasonable conjecture that, since we set $b_0 = 1$,

$$b_k = c_k, \quad k \geq 0 \tag{1}$$

and, second, that we want to find a better way to do the counting.

In fact, it turns out that there is a neat way of calculating b_k. For we see, by considering the *last* pair of parentheses, that

$$b_k = \sum_{i+j=k-1} b_i b_j, \quad k \geq 1 \tag{2}$$

For the last pair of parentheses must enclose two expressions, one involving $i + 1$ symbols and the other involving $j + 1$ symbols, for some numbers i, j satisfying $i + 1 + j + 1 = k + 1$, or $i + j = k - 1$.

Of course, we can use (2) to calculate b_k for $k = 1, 2, 3, \cdots$. For example,

$$b_3 = b_0 b_2 + b_1^2 + b_2 b_0 = 2 + 1 + 2 = 5$$

as claimed. However, our more ambitious plan is to use (2) to find a *formula* for b_k, for all $k \geq 0$.

To this end, we construct the *generating function* $\sum_{k=0}^{\infty} b_k x^k$. Then if we write $f(x)$ for this function, the recurrence relation (2) tells us that

$$f(x)^2 = \sum_{k=0}^{\infty} \left(\sum_{i+j=k} b_i b_j \right) x^k = \sum_{k=0}^{\infty} b_{k+1} x^k$$

It follows that

$$x f(x)^2 = \sum_{k=1}^{\infty} b_k x^k = f(x) - 1, \quad \text{or} \quad x f^2 - f + 1 = 0$$

where we write f for $f(x)$. Solving for f gives us $f = \frac{1 \pm \sqrt{1-4x}}{2x}$. But notice that $\frac{1+\sqrt{1-4x}}{2x} \to \infty$ as $x \to 0$. Since f is a power series in x and $f(0) = 1$, it must be the case that

$$f = \frac{1 - \sqrt{1 - 4x}}{2x} \tag{3}$$

We now expand the RHS (right-hand side) of (3) as a power series in x. We obtain[3]

$$\sqrt{1 - 4x} = (1 - 4x)^{\frac{1}{2}}$$

$$= 1 - 2x - \frac{4x^2}{2!} - \cdots - \frac{1 \cdot 3 \cdots (2k-1) 2^{k+1}}{(k+1)!} x^{k+1} - \cdots$$

so

$$f = \frac{1 - \sqrt{1 - 4x}}{2x}$$

$$= 1 + \frac{2x}{2!} + \cdots + \frac{1 \cdot 3 \cdots (2k-1) 2^k}{(k+1)!} x^k + \cdots$$

However, since $f = \sum_{k=0}^{\infty} b_k x^k$, we conclude, by equating coefficients, that

$$b_k = \frac{1 \cdot 3 \cdots (2k-1)}{(k+1)!} 2^k$$

But $\frac{1 \cdot 3 \cdots (2k-1)}{(k+1)!} 2^k = \frac{(2k)!}{(k+1)!(2 \cdot 4 \cdots 2k)} 2^k = \frac{(2k)!}{(k+1)!k!} = \frac{1}{k+1} \binom{2k}{k}$, so that

$$b_k = \frac{1}{k+1} \binom{2k}{k} \tag{4}$$

[3]Our series $\sum_{k=0}^{\infty} b_k x^k$ is really a formal power series, but the power series expansion of $(1 - 4x)^{\frac{1}{2}}$ is, in fact, valid for any real number x such that $|x| < \frac{1}{4}$. See Section 4, item (21).

Notice that (4) even holds for $k = 0$; on the other hand, an alternative form, often used and quoted, is

$$b_k = \frac{1}{k}\binom{2k}{k-1} \qquad (5)$$

which only holds if $k \geq 1$.

We move now to a third interpretation of the classical Catalan numbers. We consider **binary trees**. A tree is a graph with no circuits; and the tree is said to be binary if two branches emerge from each source node. The nodes from which no branches emerge (downwards!) are said to be **end nodes**. Our trees are *rooted* (though we draw the root at the top!), so that each edge has a preferred direction, away from the root. Then

- $a_0 = 1; a_k, k \geq 1$, is defined to be the number of rooted binary trees with k source nodes (such a tree has $(k + 1)$ end nodes, thus $(2k + 1)$ nodes altogether). See Figure 2 for examples of binary trees.

Two trees are regarded as equivalent if and only if one may be obtained from the other by pairing off nodes and (directed) edges so that the incidence relation of an edge having a node is preserved; and left and right are

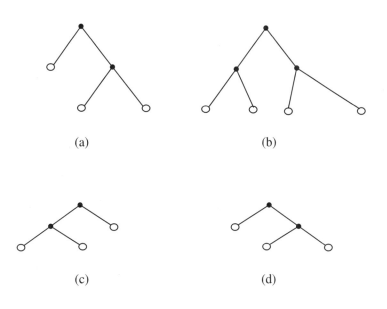

(a)   (b)

(c)   (d)

FIGURE 2

preserved. Thus Figure 2(a) and Figure 2(c) are considered different trees, but Figure 2(a) and Figure 2(d) are considered to be the same tree.

● ● ● BREAK 3

(1) Draw all the binary trees with 3 source nodes to determine the value of a_3.
(2) Draw all the binary trees with 4 source nodes to determine the value of a_4.
(3) Compare your answers to those you obtained for b_k and c_k.

It turns out to be relatively simple to see that

$$a_k = b_k \tag{6}$$

Indeed, Figure 3 suggests how to associate rooted binary trees and parenthesized expressions in one-to-one correspondence by illustrating a particular but not special case.

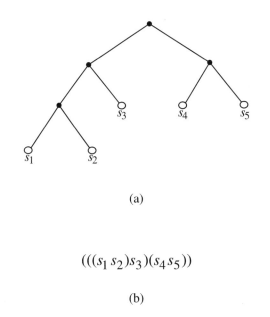

(a)

$$(((s_1 \, s_2)s_3)(s_4 \, s_5))$$

(b)

FIGURE 3　The binary tree (a) corresponding to the parenthesized expression (b).

• • • **BREAK 4**

Try to give a description of this one-to-one correspondence.

We close this introduction by sketching an argument that shows that

$$a_k = c_k \tag{7}$$

Of course, we then also know that (1) is true, so that

$$a_k = b_k = c_k = \frac{1}{k+1}\binom{2k}{k}, \quad k \geq 0 \tag{8}$$

We now illustrate how to prove (7) by means of a particular but not special case. Let us consider the convex hexagon of Figure 4(a) divided into 4 triangles by means of 3 non-intersecting diagonals.

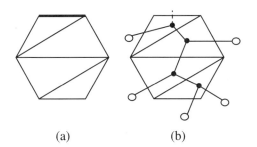

(a)   (b)

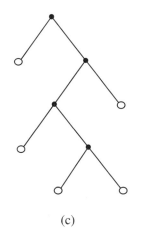

(c)

FIGURE 4

We orient our hexagon with the topmost edge horizontal, and we enter the hexagon from the outside by crossing that edge. This establishes a source node in one of the triangles. We then leave the triangle by either of the other two edges of that triangle. If we again reach the outside, we set up an end node there. If we enter a second triangle of the subdivided hexagon, we establish a source node there and repeat the process, carrying on as long as we can. The result is Figure 4(b). Figure 4(c) shows how the rooted tree of Figure 4(b) may be represented in our more usual format, but it is obvious that these are the same rooted trees.

Thus we have associated a rooted binary tree with our subdivided hexagon. Conversely, given any rooted binary tree with 5 end nodes, we construct a convex hexagon by making sure that all the end nodes are outside and that the initial node is inside (of course, we may have to distort the tree in order to construct the hexagon with straight sides; then the diagonals are inserted to cross the branches of the tree.

We will now take you through a fairly straightforward argument to show that $b_k = c_k$. Again, we will look at a particular but not special case. Figure 5(a) shows a dissected hexagon. Think of the top side as preferred and label the remaining sides in order in a counterclockwise direction as shown. Then think in terms of vector addition. The only triangle where two sides are known has sides labeled 3 and 4. Label the other side of that triangle (34). Then the next triangle where two sides are known involves sides labeled 2 and (34). Label the third side of this triangle (2(34)). Continue this process as shown in Figure 5(a) until you have the top side labeled; the label will be (1(2(34))5). This tells you how the parentheses should go in the expression involving $s_1s_2s_3s_4s_5$, namely, $(s_1(s_2(s_3s_4))s_5)$.

Going in the reverse direction involves beginning with a polygon with one more side than the number of symbols you are parenthesizing. For example, to convert the parenthesized symbol $((s_1s_2)((s_3s_4)s_5))$ into a dissected polygon you first observe that there are 5 symbols and draw a regular hexagon, labeling its sides as shown in the first part of Figure 5(b). Then, suppressing the s's and looking just at the subscripts $((12)((34)5))$, you start with any pair of parentheses that encloses just two digits, that is, in our case, (12) and (34), and draw the lines that connect their endpoints (both diagonals in this case) as shown by the dotted lines in the second part of Figure 5(b). Label these diagonals as shown in the third part of the figure. At this point there is only one triangle with two sides labeled; it has sides labeled (34) and 5. Connect the ends of those line segments as shown in the fourth part of Figure 5(b) and label it ((34)5) as shown in the last part of Figure 5(b). You then know two sides of the triangle whose third side

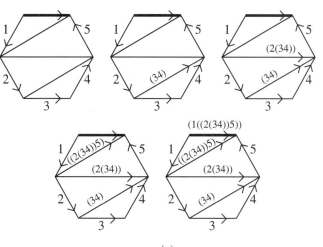

(a)

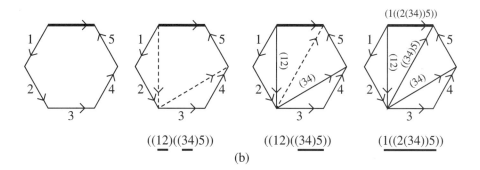

(b)

FIGURE 5

is the top edge, so you may label the top edge ((12)((34)5)). The label for the top edge will always reproduce the parenthesized expression.

● ● ● **BREAK 5**

(1) Start with the parenthesized expression $(s_1(s_2(s_3s_4))s_5)$ and carry out this procedure to show that you get the dissected polygon shown in the first part of Figure 5(a).

FIGURE 6

(2) Starting with the dissected polygon in the last part of Figure 5(b) with labels only on its sides, show that you get the parenthesized expression $((s_1 s_2)((s_3 s_4) s_5))$.

(3) What is the parenthesized expression associated with the dissected polygon shown in Figure 6?

(4) Draw the dissected 7-gon associated with the parenthesized expression $(((s_1 s_2)(s_3 s_4))(s_5 s_6))$.

(5) Draw the tree associated with the parenthesized expression of part (4).

7.2 A FOURTH INTERPRETATION

We now give a fourth interpretation of the classical Catalan numbers which plays a role in the sequel (see [5, 6]). We define

- $d_0 = 1$; d_k, $k \geq 1$, is the number of paths on the integral lattice in the plane, proceeding from the point $(0, -1)$ to the point $(k, k-1)$, that always stay below the line $y = x$.

Let us explain what this means. The *integral lattice in the plane* is the collection of points (x, y) in the coordinate plane such that x and y are both integers. A **path** on the integral lattice from $P(a, b)$ to $Q(c, d)$, where $a \leq c$, $b \leq d$, is a sequence of points $P_0 P_1 \cdots P_n$, $n \geq 0$, such that $P_0 = P$, $P_n = Q$, and P_{i+1} is obtained from P_i by moving one step to the right (east) or one step up (north), $0 \leq i \leq n - 1$. Thus if $P_i = (x_i, y_i)$, then $P_{i+1} = (x_i + 1, y_i)$ or $(x_i, y_i + 1)$. Then d_k is the number of such paths from $(0, -1)$ to $(k, k-1)$ that never meet the line $y = x$. It will be useful to designate such paths as *good*, and to refer to those paths that do meet the line $y = x$ as *bad*.[4]

Obviously, $d_0 = 1$; obviously, too, if $k \geq 1$, then a good path from $(0, -1)$ to $(k, k - 1)$ must proceed first to $(1, -1)$ (why?), so, for $k \geq 1$, we may replace $(0, -1)$ by $(1, -1)$ in the definition of d_k. It is also obvious that the penultimate stop on a good path to $(k, k - 1)$ must be at $(k, k - 2)$, so we may replace $(k, k - 1)$ by $(k, k - 2)$ in the definition of d_k.

How should we count the number of good paths? In fact, it turns out to be just as easy to count the number of good paths between *any* two integral points (a, b) and (c, d). First, we note that, to ensure that there are any good paths at all, we require

$$a > b, \quad c > d, \quad a \leq c, \quad b \leq d \tag{9}$$

Second, we observe that it is easy to count *all* the paths from (a, b) to (c, d). For it is easy to see that the number of paths from $(0, 0)$ to (c, d) is just the binomial coefficient[5] $\binom{c+d}{c}$. Thus the number of paths from (a, b) to (c, d), which must be the same as the number of paths from $(0, 0)$ to $(c - a, d - b)$, is

$$\binom{(c+d) - (a+b)}{c - a} \quad \text{or} \quad \binom{(c+d) - (a+b)}{c - a \quad d - b} \tag{10}$$

It follows that we can count the good paths if we can count the bad paths. Now a bad path from $P(a, b)$ to $Q(c, d)$ must meet the line $y = x$, for the first time, at some point R. Let us reflect the portion PR of the path PRQ in the line $y = x$, obtaining a path $\overline{PR}$. Then $\overline{P}RQ$ is a path from $\overline{P}(b, a)$ to $Q(c, d)$. Moreover, *every* path from $\overline{P}$ to Q must cross the line $y = x$ somewhere, since $\overline{P}$ is above the line and Q below it. Hence this reflection procedure sets up a one-to-one correspondence between bad paths from P to Q and paths from $\overline{P}$ to Q. It thus follows from formula (10) that the number of bad paths from P to Q is

$$\binom{(c+d) - (a+b)}{c - b}$$

---

[4] An interesting history of the *classical* Catalan numbers is given in [7]. But notice that their interpretation by means of lattice paths is not quite the same as ours.

[5] We often like to write $\binom{n}{r\ s}$, where $r + s = n$, instead of $\binom{n}{r}$, especially to bring out the symmetry. Thus $\binom{c+d}{c}$ would be written $\binom{c+d}{c\ d}$. See Chapter 6.

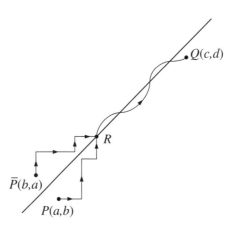

FIGURE 7   André Reflection Method.

so that the number of good paths from $P(a, b)$ to $Q(c, d)$ is

$$\binom{(c+d) - (a+b)}{c - a} - \binom{(c+d) - (a+b)}{c - b} \qquad (11)$$

● ● ●   **BREAK 6**

Calculate the number of paths from $(3, 1)$ to $(7, 6)$ that never cross the line $y = x$, though they may touch it. [Hint: What happens to a good path from $(4, 1)$ to $(8, 6)$ if you shift it one place to the left?] Calculate the number of paths from $(3, 1)$ to $(7, 6)$ that touch the line $y = x$ without crossing it.

Our method of calculating the number of good paths by means of (11) is called the André Reflection Method, discovered in the late 19th century by the Belgian-born mathematician Désiré André. We will describe the context in which he discovered the method at the end of this section, but first we use (11) to calculate d_k, the number of good paths from $(0, -1)$ to $(k, k - 1)$; according to (11) we have

$$d_k = \binom{2k}{k} - \binom{2k}{k + 1} = \binom{2k}{k}\left(1 - \frac{k}{k + 1}\right) = \frac{1}{k + 1}\binom{2k}{k} \qquad (12)$$

thus obtaining the claimed equality

$$a_k = b_k = c_k = d_k = \frac{1}{k+1}\binom{2k}{k} \tag{13}$$

We will now show you how to set up a one-to-one correspondence[6] between the set of parenthesized expressions of $s_1 s_2 \cdots s_{k+1}$ and the set of good paths from $(0, -1)$ to $(k, k-1)$. Such a parenthesized expression is a sequence of symbols, where each symbol is (i) a left parenthesis, (ii) an element s, or (iii) a right parenthesis. We first point out an interesting fact.

Proposition 1   *The right parentheses are logically superfluous in a parenthesized expression.*

Proof   We prove this by induction on k. If $k = 1$, we certainly don't need the right parenthesis to attach a unique meaning to $(s_1 s_2)$. Now suppose we don't need the right parentheses when we make k applications of our binary operation, and suppose we have an expression

$$(\cdots \ (\cdots \ (\cdots$$

involving $(k + 1)$ applications of the operation, in which we have only inserted the (correct) left parentheses, not the right parentheses. We look along the expression, from left to right, until we reach our *last* left parenthesis (it is, of course, the $(k + 1)$st). It must be followed by two element symbols, say $s_i s_{i+1}$ and, in principle, a right parenthesis. We replace $(s_i s_{i+1}$ by s and thus produce an expression involving only k applications of our binary operation and suitable left parentheses, which therefore, by the inductive hypothesis, has a unique meaning.   □

Armed with this proposition, we represent a parenthesized expression as a sequence of (i) left parentheses, and (ii) s-symbols. We now start at $(0, -1)$ as our initial integral lattice point and interpret a left parenthesis as an instruction to move right (east) and an s-symbol as an instruction to move up (north); notice that the final term of our parenthesized expression must by an s-symbol — so, too, must the penultimate term. There are $(k + 1)$ s-symbols and k left parentheses, so our path takes us from $(0, -1)$ to (k, k);

---

[6]The proved relation (13) does not make this superfluous, since it is necessary to set up a similar correspondence when we come to study, in the next section, Catalan numbers in a broader sense.

but it terminates by visiting $(k, k-2)$, $(k, k-1)$, (k, k), so we may regard it as a path from $(0, -1)$ to $(k, k-1)$. *We claim it is a good path.*

This, again, we must prove by induction on k. If $k = 1$, our parenthesized expression is $(s_1 s_2$, yielding the path

$$(0, -1) \rightarrow (1, -1) \rightarrow (1, 0) \rightarrow (1, 1)$$

which (ignoring the last step) is obviously good. To carry out the inductive step, suppose a parenthesized expression involving k applications corresponds to a good path, and suppose we are given a parenthesized expression involving $(k + 1)$ applications. Let $\cdots (s_i s_{i+1} \cdots$ mark the last appearance of a left parenthesis. If we replace $(s_i s_{i+1}$ by s we get an expression involving k applications,[7] which thus corresponds to a good path, except that it ends at (k, k). There are now 2 possibilities. If the k-expression ended in s, then the corresponding path ended in (k, k), so its penultimate vertex was $(k, k-1)$, and thus our $(k + 1)$-expression corresponds to a path that was good up to $(k, k-1)$ and then continued

$$(k, k-1) \rightarrow (k+1, k-1) \rightarrow (k+1, k) \rightarrow (k+1, k+1)$$

and thus stayed good. If the k-expression does not end in s, then the path corresponding to the k-expression as far as s is entirely below the line $y = x$, with s corresponding to some point (u, v), with $u > v$. Thus in our original $(k + 1)$-expression we move

$$(u, v-1) \rightarrow (u+1, v-1) \rightarrow (u+1, v) \rightarrow (u+1, v+1)$$

and then continue as for the k-expression except that both coordinates are increased by 1. This obviously creates a good path. This completes the inductive step.

It is not hard to believe that, conversely, every good path from $(0, -1)$ to $(k, k-1)$, completed by the vertical step $(k, k-1) \rightarrow (k, k)$, corresponds to a meaningful sequence of left parentheses and s-symbols. Try it with some examples if you don't feel you could formulate a precise proof by induction on k. Such a proof would then show that $b_k = d_k$ without invoking any calculation of either.

We close this section by describing, as promised, the context in which André first exploited his *reflection method*. It was to give a solution of the celebrated *Ballot Problem*, which had exercised the minds of many experts in probability theory in the late 19th century. In fact, a solution of the problem had been given by a French mathematician, Joseph Louis François Bertrand, but his method of solution was very complicated and provided little insight. André wrote a paper [1] entitled "Solution directe

---

[7]For brevity we call the expression a *k-expression*.

du problème resolu par M. Bertrand" (Direct solution of the problem solved by M. Bertrand), which, by contrast, gave a wonderfully concise and insightful — though we would not say "direct" — solution to the Ballot Problem.

What, then, was the Ballot Problem? We suppose there is an election with just two candidates, with the uninteresting names X and Y. We suppose further that, in fact, X wins the election with a votes against b votes for candidate Y. There is just one teller who counts the votes, one by one, and who records the number of votes for X and Y after each vote is counted. The Ballot Problem seeks an answer to the question, "What is the probability that, throughout the counting of the votes, X is always ahead of Y?"[8]

André represented the process of counting the votes as a path, in our sense, from $(0, 0)$ to (a, b). Since X won, we know that $a > b$, so that (a, b) is below the line $y = x$, and the Ballot Problem simply asks for the probability that the path from $O(0, 0)$ to $Q(a, b)$ is, except at its initial point O, a good path (see Figure 8). Here, in its essence, is André's solution.

Let us consider a path ℓ from $(0, 0)$ to (a, b).

Let $p = \text{prob}(\ell \text{ is a good path})$

$q = \text{prob}(\ell \text{ is a bad path})$

$q_N = \text{prob}(\ell \text{ is a bad path and starts north})$

$q_E = \text{prob}(\ell \text{ is a bad path and starts east})$

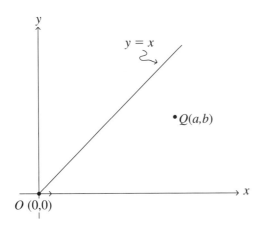

FIGURE 8

[8]This is really a general type of problem in sampling theory.

Obviously, $p + q = 1$, $q_N + q_E = q$. Also, if ℓ starts north, it is *certainly* a bad path, so $q_N = \text{prob}(\ell \text{ starts north}) = \frac{b}{a+b}$, since ℓ takes b steps north and a steps east. Now we use the André reflection method to show that $q_N = q_E$. For, given a bad path ℓ starting east, let R be the first place (after O) where ℓ meets the line $y = x$. Thus we may regard ℓ as the union $\ell' \cup \ell''$, where ℓ' is the part from O to R and ℓ'' is the part from R to Q. We reflect ℓ' in the line $y = x$ to get $\tilde{\ell}'$ and let $\tilde{\ell} = \tilde{\ell}' \cup \ell''$. Then $\tilde{\ell}$ is a path from O to Q that starts north. Conversely, we may, in a similar way, associate, with every path from O to Q starting north, a bad path from O to Q starting east. Thus, as claimed, $q_N = q_E$.

The rest is easy: $q = q_N + q_E = 2q_N = \frac{2b}{a+b}$, so $p = 1 - q = \frac{a-b}{a+b}$. We hope you agree that the method of proof is quite beautiful, but notice that the solution has some surprising features. Prominent among them is the conclusion that the required probability p depends only on the ratio of a to b. If there are 100 votes, and X gets 60 while Y gets 40, the probability p is $\frac{1}{5}$. If there are 1000 votes and X gets 600 while Y gets 400, the probability p is still $\frac{1}{5}$.

● ● ●   **BREAK 7**

(1) Can we conclude that the probability p is unchanged if we only record the situation after each batch of 10 votes has been counted?

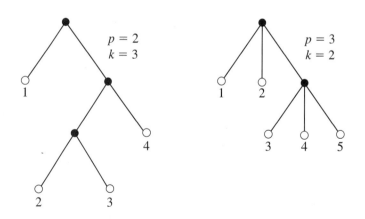

FIGURE 9   At left, a binary (or 2-ary) tree with 3 source nodes (•) and 4 endnodes (○) and, at right, a ternary (or 3-ary) tree with 2 source nodes (•) and 5 endnodes (○).

(2) Count the good paths from O to $Q(12, 5)$. Count all paths from O to $Q(12, 5)$ and thus give a somewhat different proof of the Bertrand–André result.
(3) Calculate d_k up to $k = 10$.
(4) Calculate c_k up to $k = 10$.

7.3 CATALAN NUMBERS

You may well have been wondering why we have insisted till now on referring to the *classical* Catalan numbers. The reason is that it is now standard practice to refer to the Catalan numbers as numbers (in any of the interpretations we have given you) that depend not only on a parameter k which is a non-negative integer but also on a second parameter p which is a positive integer ≥ 2, such that the classical Catalan numbers are given by taking $p = 2$. We will write the "p" as a subscript *preceding* the letter a, b, c, or d, omitting it only if it is safe to do so — that is, if there is no fear of ambiguity or misunderstanding. We will always have

$$_pa_0 = {_pb_0} = {_pc_0} = {_pd_0} = 1 \tag{14}$$

Then

- $_pa_k$ is the number of rooted p-ary trees with k source nodes (see Figure 9);
- $_pb_k$ is the number of ways of introducing parentheses into an expression involving k applications of a p-ary operation (see Figure 10);
- $_pc_k$ is the number of ways of subdividing a convex polygon into $(p+1)$-gons by means of k non-intersecting diagonals (see Figure 11);
- $_pd_k$ is the number of paths on the integral lattice in the plane from $(0, -1)$ to the point $(k, (p-1)k - 1)$ that stay below the line[9] $y = (p-1)x$ (see Figure 12).

$$
\begin{array}{cc}
p = 2 & p = 3 \\
k = 3 & k = 2 \\
(s_1((s_2\,s_3)s_4)) & (s_1\,s_2\,(s_3\,s_4\,s_5))
\end{array}
$$

FIGURE 10   At left, an expression involving 3 applications of a binary operation applied to 4 symbols; and, at right, an expression involving 2 applications of a ternary operation applied to 5 symbols.

[9]We call a path that stays below the line $y = (p-1)x$ **p-good**.

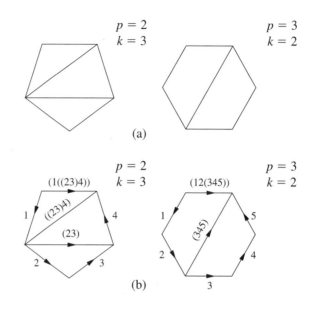

FIGURE 11   (a) At left, a 5-gon subdivided into three 3-gons by 2 diagonals and, at right, a 6-gon subdivided into two 4-gons by 1 diagonal. (b) The dissected polygons of Figure 11(a), with diagonals and last side labeled.

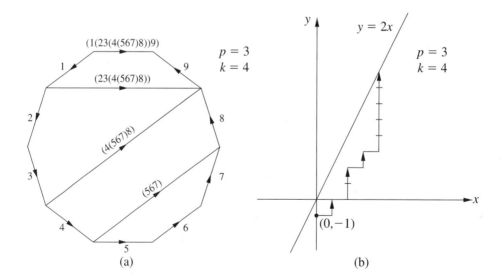

FIGURE 12   (a) The polygonal dissection associated with the expression $(s_1(s_2s_3(s_4(s_5s_6s_7)s_8))s_9)$. (b) The path associated with the expression $(s_1(s_2s_3(s_4(s_5s_6s_7)s_8))s_9)$

• • • **BREAK 8**

(1) How many nodes does a rooted p-ary tree with k source nodes possess?
(2) How long is the string $s_1 s_2 s_3 \cdots$ such that it may be parenthesized to yield k applications of a p-ary operation?
(3) How many vertices must a convex polygon have so that it can be subdivided into k polygons, each having $(p+1)$ sides, by non-intersecting diagonals? How many diagonals will be needed?

That the four interpretations of the Catalan numbers are equivalent, that is, that

$$_p a_k = {}_p b_k = {}_p c_k = {}_p d_k \tag{15}$$

is proved just as in the special case $p = 2$; if you need further clarification on this point, consult [5]. However, it turns out to be a very different matter to calculate the Catalan numbers for a general value of p. Recall that we used two methods if $p = 2$. First we calculated b_k from the recurrence relation (with $p = 2$),

$$b_k = \sum_{i+j=k-1} b_i b_j \quad (b_k = {}_2 b_k)$$

Now there is an analogue of this relation for general p, namely,

$$b_k = \sum_{i_1 + i_2 + \cdots + i_p = k-1} b_{i_1} b_{i_2} \cdots b_{i_p} \quad (b_k = {}_p b_k) \tag{16}$$

From (16) we may calculate successive values of b_k, knowing that $b_0 = 1, b_2 = p, b_3 = \frac{p(p-1)}{2}, \cdots$. However, if we want to use our generating function method to find a formula for $_p b_k$, we set $f(x) = \sum_{k=0}^{\infty} b_k x^k$. Then (16) tells us that $f^p = \sum_{k=0}^{\infty} b_{k+1} x^k$, so that

$$x f^p - f + 1 = 0 \tag{17}$$

Now, however, by contrast with the case $p = 2$, we have an equation we cannot solve by elementary methods. There is a very sophisticated method, known as the Bürmann–Lagrange inversion formula (see [5]), to obtain a power series for f from (17), but it would take us quite beyond the scope and level of this book. So we turn to our second method of calculating the classical Catalan numbers, given by $p = 2$ — using the interpretation d_k and the André reflection method. But now we want to reflect in the line $y = (p-1)x$; and it is not difficult to see that the reflection of $(0, -1)$ in this line is *not* a lattice point if $p \geq 3$. Moreover, a path on the integral lattice from a lattice point above $y = (p-1)x$ to a lattice point below $y = (p-1)x$ need not cross the line $y = (p-1)x$ at a lattice point. So the André reflection method does not extend beyond $p = 2$, and we are in real trouble!

However, we do want to calculate the Catalan numbers by a method understandable by undergraduate students. We will describe such a method in Section 5, but first we describe the idea we are going to use.

7.4 EXTENDING THE BINOMIAL COEFFICIENTS

We first meet the *binomial theorem* in elementary mathematics when we wish to expand the expression $(a+b)^n$ as a polynomial in a and b; here n is an arbitrary non-negative integer. Thus $(a+b)^n = \sum_{r=0}^{n} \binom{n}{r} a^r b^{n-r}$, where

$$\binom{n}{r} = \frac{n!}{r!(n-r)!} = \frac{n(n-1)\cdots(n-r+1)}{r!} \qquad (18)$$

The second form of (18) does not apply if $n = 0$; the first does if we interpret the factorial $0!$ as 1. We prefer here simply to specify the binomial coefficient $\binom{n}{r}$, where n is a non-negative integer, and r is a non-negative integer such that $0 \le r \le n$, by the rule

$$\binom{n}{r} = \begin{cases} 1 & \text{if } r = 0 \\ \dfrac{n(n-1)\cdots(n-r+1)}{r!} & \text{if } r > 0 \end{cases} \qquad (19)$$

We can make a further improvement, however; for it is plain from the interpretation of the binomial coefficients in terms of the expansion of $(a+b)^n$ that we would want to define $\binom{n}{r} = 0$ if $r > n$. Thus we have $\binom{n}{r}$ defined for *any n, r non-negative integers* by the same rule

$$\binom{n}{r} = \begin{cases} 1 & \text{if } r = 0 \\ \dfrac{n(n-1)\cdots(n-r+1)}{r!} & \text{if } r > 0 \end{cases} \qquad (20)$$

So far, we have only been applying the binomial theorem in its most elementary form. Now, however, we take a giant step into the unknown! We allow n to be any real number and *define the generalized binomial coefficient* for any non-negative integer r by the rule (20).

In fact, it turns out that the binomial theorem does itself generalize. Precisely, with our extended definition (20) of the binomial coefficients, it is a theorem provable by the differential calculus, that, if $|a| < |b|$, then,[10] *for any real number n,*

$$(a + b)^n = \sum_{r=0}^{\infty} \binom{n}{r} a^r b^{n-r}$$

where, as before,

$$\binom{n}{r} = \begin{cases} 1 & \text{if } r = 0 \\ \dfrac{n(n-1)\cdots(n-r+1)}{r!} & \text{if } r > 0 \end{cases} \qquad (21)$$

Actually, we will not need the full force of this theorem here.[11] We will need to know that there is a polynomial $\binom{n}{r}$ in the real variable n, of degree r, with rational coefficients, that coincides with the binomial coefficient $\binom{n}{r}$ when n is a non-negative integer. It will also be useful to us to evaluate $\binom{n}{r}$ when n is a *negative* integer. Of course,

$$\binom{n}{0} = 1 \quad \text{if } n \text{ is a negative integer.} \qquad (22)$$

Now let us suppose $r \geq 1$ and let $n = -m$, where m is a positive integer. Then

$$\binom{-m}{r} = \frac{(-m)(-m-1)\cdots(-m-r+1)}{r!}$$

$$= (-1)^r \frac{(m+r-1)(m+r-2)\cdots(m+1)m}{r!}, \quad \text{so that}$$

$$\binom{-m}{r} = (-1)^r \binom{m+r-1}{r} \qquad (23)$$

Notice that, in (23), $\binom{m+r-1}{r}$ is an ordinary binomial coefficient, since $m + r - 1 > 0$; indeed, $m + r - 1 \geq r$. We will certainly be making use of formula (23) — see Section 7.

With this (probably very unexpected!) preparation, we are ready to calculate $_p d_k$.

---

[10] We invoked this in Section 1 in our expansion of $\sqrt{1 - 4x}$.

[11] Here we are repeating some arguments made in Chapter 6 and [4], so that the reader may have a self-contained treatment.

● ● ● **BREAK 9**

(1) Use the formula in (21) to calculate $\binom{n}{r}$ when $n = -\frac{1}{2}$.
(2) Is the Pascal Identity $\binom{n}{r} + \binom{n}{r-1} = \binom{n+1}{r}$, $r \geq 1$, still true if n is a negative integer? [Hint: Use formula (23).]

7.5 CALCULATING GENERALIZED CATALAN NUMBERS

Indeed, we are ready to calculate a *generalization* of $_p d_k$; since we will hold p fixed throughout this section, we will suppress p from the notation and simply write d_k. Thus we know that $d_0 = 1$, and we wish to calculate the number of good paths from $(0, -1)$ to $(k, (p - 1)k - 1)$, $k \geq 1$. As remarked in the case $p = 2$, this is equivalent to calculating the number of good paths from $(1, -1)$ to $(k, (p - 1)k - 1)$; we designate this latter point P_k and call the point immediately above it, on the line $y = (p - 1)x$, Q_k. We will eventually want to calculate the number of good paths between *any* two points below the line $y = (p - 1)x$, and to do so we need to generalize our problem now and seek to calculate d_{qk}, the number of good paths from the integral lattice point $(1, q - 1)$ to P_k, so that $d_k = d_{0k}$ (see Figure 13). Of course, we insist that $q < p$, so that $(1, q - 1)$ is *below* the line $y = (p - 1)x$; but we allow q to take any integer value subject to this condition. It turns out to be no more difficult to calculate d_{qk} than

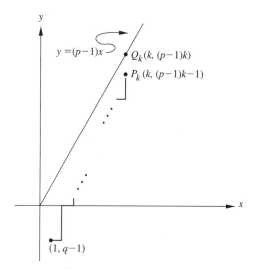

FIGURE 13

to calculate d_k. But we have to face the fact that we cannot use the André Reflection Method if $p > 2$.

We first obtain what we call a (generalized) **Jonah formula**, since the mathematician David Jonah was the first to obtain a special case of this formula. Consider a lattice point $N(k, n - k)$ *above* (or *on*) the line $y = (p - 1)x$. Then we may partition the set of paths from $(1, q - 1)$ to N according to where they first meet the line $y = (p - 1)x$. Consider the set Γ_i of paths from $(1, q - 1)$ to N that first meet $y = (p - 1)x$ at the point $Q_i(i, (p - 1)i)$. Then the number of paths in Γ_i is obviously given by the product $d_{qi} \binom{n-pi}{k-i}$; for d_{qi} is the number of good paths from $(1, q - 1)$ to $P_i(i, (p - 1)i - 1)$ and $\binom{n-pi}{k-i}$ is the number of paths from Q_i to N (see Figure 14). Moreover, plainly, $1 \le i \le k$. Since there are $\binom{n-q}{k-1}$ paths from $(1, q - 1)$ to N, we arrive at the key formula

$$\binom{n-q}{k-1} = \sum_{i=1}^{k} d_{qi} \binom{n-pi}{k-i}. \tag{24}$$

We propose to use formula (24) to calculate d_{qk} — how can we do this? It is here that we exploit the point of view of Section 4. For we must emphasize that formula (24) has so far only been established when $N(k, n - k)$ lies *above* (or *on*) the line $y = (p - 1)x$, that is, when $n \ge pk$. However, by

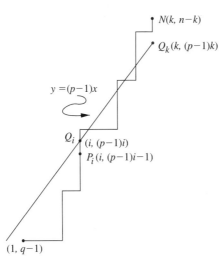

FIGURE 14 Partitioning of paths from $(1, q - 1)$ to $N(k, n - k)$ according to where they first meet the line $y = (p - 1)x$.

generalizing the meaning of $\binom{n}{r}$ by means of formula (21), to allow n to be *any* real number, (24) becomes an equality between two polynomials in the indeterminate n, with rational coefficients, which we know holds for infinitely many values of n, namely, for all integers $n \geq pk$. But then the two polynomials must be *identically equal*, so that, in fact, equality (24) must hold for *all* real values of n, and, in particular, for all *integer* values of n. In fact, we will only be substituting integers for n, but we will often substitute for n an integer $< pk$.

To sum up, we now know that we can safely substitute *any* integer for n into (24) and not only the integers n satisfying $n \geq pk$.

We immediately exploit this new freedom by taking $n = pk - 1$. We conclude that

$$\binom{pk - q - 1}{k - 1} = \sum_{i=1}^{k} d_{qi} \binom{pk - pi - 1}{k - i} \tag{25}$$

Now, if $i = k$, then $\binom{pk-pi-1}{k-1} = \binom{-1}{0} = 1$, so that (25) may be written

$$\binom{pk - q - 1}{k - 1} = \sum_{i=1}^{k-1} d_{qi} \binom{pk - pi - 1}{k - i} + d_{qk} \tag{26}$$

We now substitute $n = pk - 1$ in (24) as before, but we also replace k by $(k - 1)$. This yields

$$\binom{pk - q - 1}{k - 2} = \sum_{i=1}^{k-1} d_{qi} \binom{pk - pi - 1}{k - i - 1} \tag{27}$$

At this stage we note that we had better suppose $k \geq 2$; but this is not troublesome, for there is no problem in calculating d_{q1}. After all, d_{q1} is the number of good paths from $(1, q - 1)$ to $P_1(1, p - 2)$, and so, clearly,

$$d_{q1} = 1 \tag{28}$$

We now observe that (26) and (27) together do enable us to calculate d_{qk}, $k \geq 2$. For we always have $\binom{n}{r} = \frac{n-r+1}{r}\binom{n}{r-1}$, $r \geq 1$, so

$$\binom{pk - pi - 1}{k - i} = \frac{(p - 1)(k - i)}{k - i} \binom{pk - pi - 1}{k - i - 1}$$

$$= (p - 1) \binom{pk - pi - 1}{k - i - 1}, \quad i \leq k - 1$$

Thus we conclude from (26) and (27) that

$$
d_{qk} = \binom{pk - q - 1}{k - 1} - (p - 1)\binom{pk - q - 1}{k - 2}
$$

$$
= \binom{pk - q}{k - 1} \left\{ \frac{pk - q - k + 1}{pk - q} - \frac{(p - 1)(k - 1)}{pk - q} \right\}
$$

$$
= \binom{pk - q}{k - 1} \frac{p - q}{pk - q}, \quad k \geq 2
$$

Finally, we have shown that

$$
d_{qk} = \frac{p - q}{pk - q} \binom{pk - q}{k - 1}, \quad k \geq 1 \tag{29}
$$

since the RHS of (29) gives the value 1 if $k = 1$; as remarked, there is only one path from $(1, q - 1)$ to $(1, p - 2)$, and it is good! Of course, we should notice that (29) gives us the correct value for d_k by setting $q = 0$, namely, $d_k = d_{0k} = \frac{1}{k}\binom{pk}{k-1}$. We announce this as a corollary.

Corollary 2 *The Catalan numbers ${}_pC_k$ are given by*

$$
{}_pC_0 = 1, \quad {}_pC_k = \frac{1}{k}\binom{pk}{k - 1}, \quad k \geq 1
$$

Here we have written C_k for the kth Catalan number, in *any* of its interpretations. It is intriguing that the formula for ${}_pC_k$ is a very natural generalization of that for the classical Catalan number ${}_2C_k$; but we were not able to generalize André's lovely reflection method. Have we missed something?

7.6 COUNTING *p*-GOOD PATHS

In this section we will justify our claim that we can now calculate the number of p-good paths between any two points (a, b) and (c, d) below the line $y = (p - 1)x$.

We first point out that we may always assume that our initial point has the form $(1, q - 1)$. For let us slide the point (a, b) along the line parallel to $y = (p - 1)x$ through (a, b) until the x-coordinate reaches the value 1. It is clear that the y-coordinate then

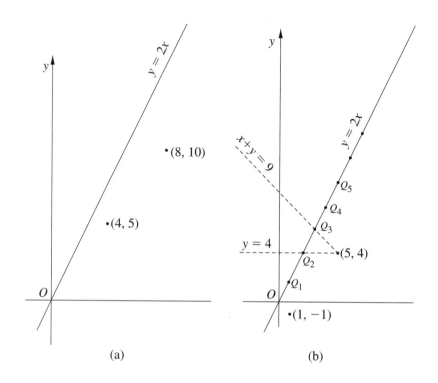

FIGURE 15 (a) Counting the 3-good paths from $(4, 5)$ to $(8, 10)$. (b) After sliding: counting the 3-good paths from $(1, -1)$ to $(5, 4)$.

reaches the value $q - 1$, where $q - 1 = b - (p - 1)(a - 1)$, that is, $q = b - (p - 1)a + p$. Moreover, the point (c, d) reaches the point $(c', d') = (c - a + 1, d - (p - 1)(a - 1))$, so that the set of good[12] paths from (a, b) to (c, d) is in one-to-one correspondence with the set of good paths from $(1, q - 1)$ to (c', d'), where q has the value above. Finally, we may, of course, write $(k, n - k)$ for (c', d'), where $k = c'$, $n = c' + d'$. Thus we have to count the good paths from $(1, q - 1)$ to $N(k, n - k)$. See Figure 15 for an example of the sliding procedure.[13]

---

[12] We suppress the symbol p.

[13] You will not understand all of Figure 15(b) until you have read further.

• • • **BREAK 10**

Choose a fixed integer r with $1 \leq r \leq p$. If $k \geq 1$ and $q + r \leq p - 1$, show that

$$d_{qk} = \sum_{\substack{i,j \geq 1 \\ i+j=k+1}} d_{p-r,i} d_{q+r,j}$$

[Hint: Partition the p-good paths from $(1, q - 1)$ to P_k according to where they first meet the line $y = (p - 1)x - r$.]

We have now put our problem in a form which allows us to exploit the key formula (24), which we repeat here for the reader's convenience:

$$\binom{n-q}{k-1} = \sum_{i=1}^{k} d_{qi} \binom{n - pi}{k - i} \qquad (30)$$

Note that the formula is still valid even if the point $(k, n - k)$ lies *below* the line $y = (p - 1)x$, *as we will henceforth assume;* and that the LHS, that is, $\binom{n-q}{k-1}$, still represents the number of paths from $(1, q - 1)$ to $(k, n - k)$. Thus we now proceed to count the bad paths from $(1, q - 1)$ to $(k, n - k)$. We claim that a bad path from $(1, q - 1)$ to $(k, n - k)$ first meets the line $y = (p - 1)x$ at a point Q_i, where $1 \leq i \leq \ell$, and ℓ is the largest integer $\leq \frac{n-k}{p-1}$; we may write this integer

$$\ell = \left[\frac{n - k}{p - 1} \right] \qquad (31)$$

Figure 16 shows the rationale for this claim — Q_i can't be further north than $N(k, n - k)$. Of course, if $\ell = 0$, this simply means that there are *no* bad paths from $(1, q - 1)$ to $N(k, n - k)$, so there are $\binom{n-q}{k-1}$ good paths.

In general, we conclude that the number of good paths form $(1, q - 1)$ to $N(k, n - k)$ is

$$\binom{n-q}{k-1} - \sum_{i=1}^{\ell} d_{qi} \binom{n - pi}{k - i}, \quad \text{where } \ell = \left[\frac{n - k}{p - 1} \right] \qquad (32)$$

Example 3   Use formula (32) to count the 3-good paths between $(5, 0)$ and $(7, 10)$.

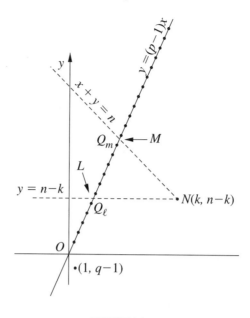

FIGURE 16

Solution   First let us simply use the formula given in the text. We have

$$a = 5, \quad b = 0, \quad p = 3 \quad \text{so} \quad q = -7, \quad c = 7, \quad d = 10,$$
$$\text{so} \quad c' = 3, \quad d' = 2, \quad k = 3, \quad n = 5,$$

and we are counting 3-good paths from $(1, -8)$ to $(3, 2)$. Notice that since $p - 1 = 2$, our rule is: if x comes down by h, then y comes down by $2h$. Since 5 goes to 1 in the slide, x comes down by 4, so y comes down by 8. Hence $(5, 0)$ slides to $(1, -8)$, and $(7, 10)$ slides to $(3, 2)$.

Our formula (32) gives us for the number of 3-good paths,

$$\binom{12}{2} - \sum_{i=1}^{1} d_{-7,i} \binom{5 - 3i}{3 - i}$$

since $\frac{n-k}{p-1} = 1$. Thus we have $\binom{12}{2} - \binom{2}{2} = 65$. (A picture makes it obvious that there is only one bad path.)

Of course, we can use (30) to obtain a second formula for the number of good paths from $(1, q - 1)$ to $(k, n - k)$, namely,

$$\sum_{i=\ell+1}^{k} d_{qi} \binom{n - pi}{k - i}, \quad \text{where } \ell = \left[\frac{n - k}{p - 1}\right] \tag{33}$$

However, we may improve on formula (33) in a rather interesting way. We remarked, following (19), that the (generalized) binomial coefficient $\binom{n}{r}$ is zero if n, r are non-negative integers with $r > n$. Now, in (33), $i \geq \ell + 1$, so $i > \frac{n-k}{p-1}$, whence $k - i > n - pi$. Thus, in the summation (33), the (generalized) binomial coefficient $\binom{n-pi}{k-i}$ is zero so long as $n - pi \geq 0$. Let us set $m = [\frac{n}{p}]$. Then $\frac{n}{p} > \frac{n-k}{p-1}$, since $(k, n - k)$ is *below* the line $y = (p - 1)x$, so $m \geq \ell$, and, as we have argued, $\binom{n-pi}{k-i} = 0$ if $\ell + 1 \leq i \leq m$. We conclude that the number of good paths from $(1, q - 1)$ to $(k, n - k)$ is

$$\sum_{i=m+1}^{k} d_{qi} \binom{n - pi}{k - i}, \quad \text{where } m = \left[\frac{n}{p}\right] \qquad (34)$$

(You should now fully understand the presence of the dashed lines on Figures 15(b) and 16.)

It is a matter of choice whether we use (32) or (34) to calculate the number of good paths. In any particular case, one will probably be more convenient than the other. Where good paths predominate, (32) is to be preferred; where bad paths predominate, (34) is to be preferred.

• • • **BREAK 11**

(1) Use (34) to compute the number of 4-good paths from $(1, 2)$ to $(4, 10)$.

(2) Show that the number of p-good paths from $(1, q - 1)$ to $(k, (p - 1)k - r)$, $1 \leq r \leq p$, is independent of r (a) by using formula (34), and (b) by means of a picture. What is the number?

7.7 A FANTASY — AND THE AWAKENING

Let us write $\Pi_g(k, n)$ for the number of p-good paths from $(1, q - 1)$ to $(k, n - k)$; notice that we suppress p and q from the notation, since we will not vary them in our discussion in this section. Thus[14]

$$\Pi_g(k, n) = \sum_{i=m+1}^{k} d_{qi} \binom{n - pi}{k - i}, \quad \text{where } m = \left[\frac{n}{p}\right] \qquad (35)$$

Let us recall, from our proof of formula (24), how we first obtained the individual terms $d_{qi} \binom{n-pi}{k-i}$. The factor d_{qi} came from counting the good

---

[14] The subscript g denotes "good."

paths from $(1, q - 1)$ to P_i, and the factor $\binom{n-pi}{k-i}$ came from counting the paths from Q_i to $N(k, n - k)$. Thus we may imagine that, to obtain a count of the good paths from $(1, q - 1)$ to N, we partition them according to the lattice points Q_i on the line $y = (p - 1)x$, $m + 1 \le i \le k$, and then the contribution from Q_i is the product $d_{qi}\binom{n-pi}{k-i}$. The formalism is right; but the interpretation is sheer fantasy! For a good path from $(1, q - 1)$ to N does *not* meet the line $y = (p - 1)x$; the binomial coefficient $\binom{n-pi}{k-i}$ *doesn't* count the number of paths from Q_i to N, since there are no paths from Q_i to N when $m+1 \le i \le k$ (since then Q_i is above N); *and* $d_{qi}\binom{n-pi}{k-i}$ *doesn't count anything*, since its values in the range $m + 1 \le i \le k$ are alternately positive and negative, ending with the positive value d_{qk} when $i = k$.

Nevertheless, formula (35) is perfectly correct, and our bizarre interpretation of it, completely consistent with our valid interpretation of the corresponding formula[15]

$$\Pi_b(k, n) = \sum_{i=1}^{\ell} d_{qi}\binom{n - pi}{k - i}, \quad \text{where } \ell = \left[\frac{n - k}{p - 1}\right] \qquad (36)$$

for the count $\Pi_b(k, n)$ of the bad paths from $(1, q - 1)$ to $N(k, n - k)$, is surely helpful, even if absurd (see Figure 17). However, we should perhaps offer a more prosaic explanation of formula (35).[16]

To do so, we recall from Section 4 the rule

$$\binom{n}{r} = (-1)^r\binom{r - n - 1}{r}, \quad \text{where } n \text{ is negative.} \qquad (37)$$

Thus we may rewrite (35) as

$$\Pi_g(k, n) = \sum_{i=m+1}^{k} (-1)^{k-i} d_{qi}\binom{(p - 1)i - n + k - 1}{k - i}, \quad \text{where } m = \left[\frac{n}{p}\right] \qquad (38)$$

We are going to interpret (38) as the formula we get from the *initial conditions*

$$\Pi_g(k, n) = d_{qk}, \quad n = pk - 1, pk - 2, \cdots, pk - p \qquad (39)$$

---

[15]The subscript b denotes "bad".

[16]We are grateful to our colleague Tamsen Whitehead for her essential contribution to this demystification. See [6].

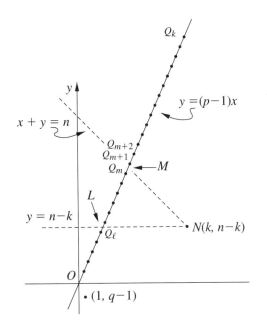

FIGURE 17

and the *recurrence relation*, valid when $n + 1 < pk$, i.e., $n < pk - 1$,

$$\Pi_g(k, n + 1) = \Pi_g(k, n) + \Pi_g(k - 1, n) \tag{40}$$

For (40) holds, since a good path from $(1, q - 1)$ to $(k, n - k + 1)$ must make its penultimate stop either at $(k, n - k)$ or at $(k - 1, n - k + 1)$ (see Figure 18).

We establish (38) by a downward induction on $n - k$, applying the recurrence relation (40) in the form

$$\Pi_g(k, n) = \Pi_g(k, n + 1) - \Pi_g(k - 1, n). \tag{41}$$

Notice that (39) and (41) explain why $\Pi_g(k, n)$ is a linear combination of the generalized Catalan numbers d_{qi}, and why the actual linear combination has alternating signs. Let us show just how we get the precise formula (38) by this argument.[17]

---

[17]We emphasize that we have already *proved* (38) by a totally different argument.

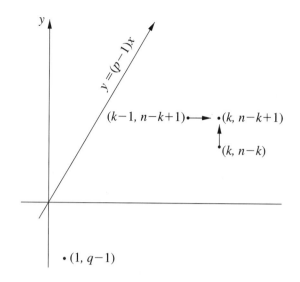

FIGURE 18 $\Pi_g(k,n)$ **represents the number of** *p***-good paths from** $(1, q - 1)$ **to the point** $N(k, n - k)$. **Thus** $\Pi_g(k, n + 1) = \Pi_g(k, n) + \Pi_g(k - 1, n)$.

Our inductive argument is that if the two terms on the right of (41) are given by (38), then so is the term on the left. Thus we substitute from (38) into the right of (41), getting

$$\sum_{i=\left[\frac{n+1}{p}\right]+1}^{k} (-1)^{k-i} d_{qi} \binom{(p-1)i - n + k - 2}{k - i}$$

$$+ \sum_{i=\left[\frac{n}{p}\right]+1}^{k-1} (-1)^{k-i} d_{qi} \binom{(p-1)i - n + k - 2}{k - 1 - i} \qquad (42)$$

and we want to show that this is just

$$\sum_{i=\left[\frac{n}{p}\right]+1}^{k} (-1)^{k-i} d_{qi} \binom{(p-1)i - n + k - 1}{k - i} \qquad (43)$$

All would be well if all summations went from $[\frac{n}{p}] + 1$ to k — we would simply apply the famous *Pascal Identity* (see Chapter 6). However, there are two apparent snags, namely,

(i)   in the first summand in (42) we start with $i = [\frac{n+1}{p}] + 1$; and

(ii)   in the second summand in (42) we end with $i = k - 1$.

To deal with (i), notice that $[\frac{n+1}{p}] + 1$ differs from $[\frac{n}{p}] + 1$ only when n has the form $n = pj - 1$, in which case we start the first summation at $i = j + 1$ instead of $i = j$. But when we substitute $i = j, n = pj - 1$ into the binomial coefficient in the first summation we get

$$\binom{(p-1)j - (pj - 1) + k - 2}{k - j} = \binom{k - j - 1}{k - j} = 0;$$

remember that $n < pk - 1$, so $j < k$.

To deal with (ii), notice that if $i = k$, the binomial coefficient in the second summand is

$$\binom{(p-1)k - n + k - 2}{-1} = \binom{(p-1)k - n + k - 2}{(p-1)k - n + k - 1} = 0;$$

here we use the symmetry condition $\binom{n}{r} = \binom{n}{n-r}$ to define $\binom{n}{r}$ when r is a negative integer, remarking that this is consistent with the Pascal Identity when $n \geq 0$, yielding in this case $\binom{n}{r} = 0$.

This completes our combinatorial interpretation of formula (38) (or (35)).

Finally, we give a geometrical background for the counting of paths (see Figure 19 for a typical example). We have the formula (with fixed p, q and $n < pk$)

$$\binom{n - q}{k - 1} = \Pi_b(k, n) + \Pi_g(k, n)$$

where $\Pi_b(k, n)$, the number of bad paths from $(1, q - 1)$ to $N(k, n - k)$, is given by

$$\Pi_b(k, n) = \sum_{i=1}^{\ell} d_{qi} \binom{n - pi}{k - i}, \quad \ell = \left[\frac{n - k}{p - 1}\right]$$

and Π_g, the number of good paths from $(1, q - 1)$ to $N(k, n - k)$, is given by

$$\Pi_g(k, n) = \sum_{i=m+1}^{k} d_{qi} \binom{n - pi}{k - i}, \quad m = \left[\frac{n}{p}\right]$$

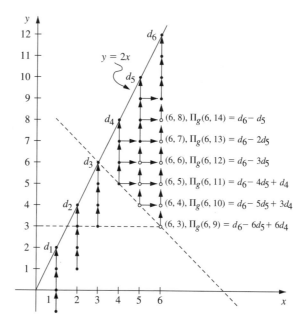

FIGURE 19   Here $p = 3, q = -1$, and we adopt the notational simplification d_i for $d_{-1,i}$. Recall that $\Pi_g(k, n)$ here represents the number of 3-good paths from $(1, q - 1)$ to the point $N(k, n - k)$.

We draw a horizontal line through N meeting the line $y = (p - 1)x$ in L, and Q_ℓ is the last lattice point on $y = (p - 1)x$ on the way to L. We then draw a line through N with slope -1 meeting the line $y = (p - 1)x$ in M, and Q_m is the last lattice point on $y = (p - 1)x$ on the way to M. The count of paths from $(1, q - 1)$ to $N(k, n - k)$ is "partitioned" by the lattice points $Q_1, Q_2, \cdots, Q_k$. The points $Q_1, Q_2, \cdots, Q_\ell$ provide a genuine partitioning of the set of bad paths; the lattice points $Q_{\ell+1}, \cdots, Q_m$ provide a zero count; and the points $Q_{m+1}, \cdots, Q_k$ provide, as explained, a pseudo-partitioning of the set of good paths. Figure 16 shows the general case; and Figure 19 shows how formula (38) actually arises from the initial conditions (39) and the recurrence formula (41).

● ● ●   **BREAK 12**

    Study Figure 19 carefully and make sure you understand what it signifies.

REFERENCES

1. André, D, Solution directe du problème résolu par M. Bertrand, *Comptes Rendus Acad. Sci. Paris* **105** (1887), 436–7.
2. Gould, Henry W., Bell and Catalan Numbers, Research bibliographies of two special number sequences, available from the author (Department of Mathematics, West Virginia University, Morgantown, WV 26506). The 1979 edition sells for $3.00 and contains over 500 references pertaining to Catalan numbers.
3. Hilton, Peter, Derek Holton, and Jean Pedersen, *Mathematical Reflections — In a Room with Many Mirrors*, 2nd printing, Springer-Verlag, NY, 1998.
4. Hilton, Peter, and Jean Pedersen, Extending the binomial coefficients to preserve symmetry and pattern, *Computers Math. Applic.* **17** (1–3) (1989), 89–102.
5. Hilton, Peter, and Jean Pedersen, Catalan numbers, their generalizations and their uses, *The Mathematical Intelligencer* **13**, No. 2 (1991), 64–75.
6. Hilton, Peter, Jean Pedersen, and Tamsen Whitehead, On paths on the integral lattice in the plane, *Far East Journal of Mathematical Sciences*, Feb. (1999), 1–23.
7. Larcombe, P.J., and P.D.C. Wilson, On the trail of the Catalan sequence, *Mathematics Today*, **35**, No. 1 (1998), 114–117.
8. Sloane, N.J.A., and Simon Plouffe, *The Encyclopedia of Integer Sequences*, New York and London, Academic Press, 1995.
9. Stanley, Richard P., *Enumerative Combinatorics*, vol. 2, Cambridge University Press, 1999.

8 | Symmetry

8.1 INTRODUCTION: A REALLY BIG IDEA

The concept of symmetry plays a strong role today in many of the exact sciences. Thus, for example, theoretical physicists, in searching for a unified field theory, have been led to the notion of *supersymmetry*, applied to the (super)strings, which, as some believe, are the fundamental building blocks of the universe. Perhaps the foremost exponent of this position is the American physicist Edward Witten, of the Princeton Institute for Advanced Study, who, a few years ago, won a Fields Medal — the most prestigious award that can be given to a mathematician[1] — for his fundamental theoretical contributions to superstring theory. Even more recently (August 1998, at the International Congress of Mathematicians held in Berlin), the Cambridge mathematician Richard Borcherds was awarded a Fields Medal for his contribution to the development of symmetry theory, especially with respect to Witten theory and its relation to the advanced mathematical theory of sporadic finite groups.

What, then, *is* symmetry? In this chapter we attempt to make the idea precise, keeping our applications of the concept at a level where, as we hope, they will be appreciated by our readers. Nevertheless, there is much

---

[1] It is a curious fact that the only mathematician to receive a Nobel Prize for mathematical work was John Nash, whose work on equilibrium had highly significant consequences in the study of economics. So John Nash was awarded a Nobel Prize as an economist; there is no Nobel Prize for mathematics.

intrinsic difficulty in the mathematics of this chapter, so do not expect to be able to understand it without careful attention to the arguments and the examples.[2]

We will confine ourselves to the use of the symmetry concept within mathematics; and we must first of all emphasize that notions of symmetry, while fundamental to geometry, are certainly, and importantly, to be found in mathematics outside geometry. Thus, for example, any function of two variables $f(x, y)$ may be described as *symmetric* if $f(x, y) = f(y, x)$; for example, $f(x, y) = x^2 + y^2$ or $f(x, y) = 5xy$. Similarly, any function of three variables $f(x, y, z)$ may be described as *symmetric* if its value remains unchanged under any permutation (there are 6 if one includes the identity permutation) of the variables x, y, z; for example, $f(x, y, z) = x^4 + y^4 + z^4 - 3xyz$. As you will have seen in Chapter 6, the binomial coefficient $\binom{n}{r}$ becomes a symmetric function if one regards it as a function of r and s, where $r + s = n$. Thus the relation $\binom{n}{r} = \binom{n}{n-r}$ takes the *symmetric* form

$$\binom{n}{r\ s} = \binom{n}{s\ r}$$

if one writes $\binom{n}{r\ s}$, with $r + s = n$, instead of the (apparently) simpler $\binom{n}{r}$. Similarly, the trinomial coefficient $\binom{n}{r\ s\ t}$, $r + s + t = n$, is a symmetric function of the variables[3] r, s, t.

As another example, symmetric functions of the roots of polynomial equations

$$x^n + a_1 x^{n-1} + \cdots + a_{n-1} x + a_n = 0$$

play an essential role in the algebra of polynomials. We know that a polynomial equation of degree n over the complex numbers has n roots. A crucial theorem states that if $\alpha_1, \alpha_2, \cdots, \alpha_n$ are the roots, then any symmetric polynomial in $\alpha_1, \alpha_2, \cdots, \alpha_n$, with integer coefficients may be expressed, uniquely, as a polynomial in $a_1, a_2, \cdots, a_n$ with integer coefficients. Thus, for example, with $n = 3$, we have

$$\alpha_1 + \alpha_2 + \alpha_3 = -a_1$$

---

[2] We advise you to read Chapter 4 before studying this chapter systematically. We have found that many students — but not all — understand the arguments about symmetry much better when they have a concrete geometrical model actually in their hands.

[3] In order to stress the symmetry, the eminent French mathematician Henri Cartan employs the simple, but revealing, notation $(r\ s)$, $(r\ s\ t)$ for binomial and trinomial coefficients, respectively. Here the n is suppressed. Why is this legitimate?

$$\alpha_1^2 + \alpha_2^2 + \alpha_3^2 = a_1^2 - 2a_2$$
$$\alpha_1^3 + \alpha_2^3 + \alpha_3^3 = -a_1^3 + 3a_1a_2 - 3a_3$$

• • • BREAK 1

(1) Check the above formulae, using the identity

$$x^3 + a_1 x^2 + a_2 x + a_3 = (x - \alpha_1)(x - \alpha_2)(x - \alpha_3)$$

(You may like to check them first in a simple particular case like $\alpha_1 = 1, \alpha_2 = 2, \alpha_3 = 3$.)

(2) Continue the sequence above by expressing $\alpha_1^4 + \alpha_2^4 + \alpha_3^4$ in terms of a_1, a_2, a_3; remember that $n = 3$.

(3) Use the crucial theorem quoted to prove that, if F_n is the nth Fibonacci number, then $F_m | F_n$ if $m | n$; and that, if L_n is the nth Lucas number, then $L_m | L_n$ if $m | n$ with odd quotient (we say that $m | n$ *oddly*). [Hint: Recall the Binet formulae from Chapter 3,

$$F_n = \frac{\alpha^n - \beta^n}{\alpha - \beta}, \quad L_n = \alpha^n + \beta^n$$

where α, β are the roots of $x^2 - x - 1 = 0$.]

However, while symmetry is valuable outside geometry, our main purpose in this chapter will, in fact, be to explain the nature and role of symmetry within geometry. It is our belief that its role in other parts of mathematics will then become easier to understand. Interestingly, it turns out that, in order to describe the nature of symmetry in geometry, it is necessary to introduce some basic concepts in a very important part of algebra known as *group theory*. Thus, in Section 2, we describe these concepts and explain how they enable us to define the key notion of the *symmetry group* of a geometric configuration A. Let us add that it is not at all surprising that we need basic concepts from group theory in geometry. For the distinguished German mathematician Felix Klein (1849–1925), in his famous manifesto, called (in German) the Klein Erlanger Programm, *defined* geometry in terms of the group of allowed transformations of a given set of points. We will be adopting Klein's point of view. For example, from this point of view, the *Euclidean Geometry* of the plane $\mathbb{R}^2$ is to be understood as the set of properties of configurations in $\mathbb{R}^2$ invariant under the group of transformations of $\mathbb{R}^2$ generated by *translations*, *rotations*, and *reflections*.

In Section 3, we use the group theory presented in Section 2 to introduce the concept of *homologue*, due to the great mathematician and expositor

George Pólya (1887–1985). Pólya never wrote this idea down, but asked two of us (PH and JP), near the end of his life, to do so on his behalf, so we take this chance to do so. In fact, our interpretation of Pólya's idea makes that idea applicable in a more general situation, namely, whenever one considers groups acting on sets — though the most vivid examples are going to come from geometry.

The precise details of our presentation of homologues are to be found at the end of Section 4, where we describe a very important theorem of combinatorics, the ***Pólya Enumeration Theorem***. This theorem is described, with justification, in the *Handbook of Applicable Mathematics* [3] as the *Redfield–Pólya Theorem*.[4] It is very often applied in a geometric context (our own applications will be geometric); however, once again, it is not in essence a theorem of geometry at all, but rests on the foundations of the idea of a (finite) group acting on a finite set — a basic idea of combinatorics.

Section 5 is a treatment of the idea of *odd* and *even* permutations. Those who would have preferred to get their group theory in one piece rather than two may, of course, read this section immediately after they have studied

---

[4]To see why, consult the article by E. Keith Lloyd, J. Howard Redfield 1879–1944, *J. Graph Theory*, **8** (1984), 195–203. There is a poem "Enumerational" in C. Berge's Principles of Combinatorics (Academic Press, 1971) that, somewhat amusingly, refers to the history of this part of mathematics.

Pólya had a theorem
(Which Redfield proved of old).
What Secrets sought by graphmen
Whereby the theorem told!

So Pólya counted finite trees
(As Redfield did before).
Their number is exactly such,
And not a seedling more.

Harary counted finite graphs
(Like Redfield, long ago).
And pointed out how very much
To Pólya's work we owe.

And Read piled graph on graph on graph
(Which is what Redfield did).
So numbering the graphic world
That nothing could be hid.

Then hail, Harary, Pólya, Read,
Who taught us graphic lore,
And spare a thought for Redfield too,
Who went too long before.

Blanche Descartes

Section 2; but we thought that, though the basic idea of the section plays an important role in the study of polyhedral symmetry, it might be rather indigestible if offered to the reader before much of the geometry had been treated.

8.2 SYMMETRY IN GEOMETRY

Although, as we have said, the concept of symmetry may be found in all parts of mathematics, and in all those areas of science to which mathematics makes an essential contribution, it remains true that its best known applications are in geometry. One finds the idea of symmetry frequently referred to in good elementary treatments of geometry (e.g., [8]), and also in good treatments of geometry suitable for those making a serious study of mathematics at the university level (e.g., [7]). But many elementary treatments of the symmetry of geometric figures are confusing and misleading, largely because those treatments never make it clear what geometry is (nor what symmetry is).

So we want to give a precise definition of *geometry*. Of course, we are not advocating that this be done the way we are doing it here if you are teaching elementary students, but it does seem to us appropriate to introduce these fundamental mathematical ideas to the readers of our book. We maintain, with Klein, that no treatment of *geometry* is complete without the idea of *symmetry*; and that no clear idea of symmetry is possible without the basic notion of a *group*.

Thus we start this section with the definition of a group.

Definition 1   Let G be a set and let $*$ be a binary operation on G; that is, given two elements g, h of G, then $g * h$ is itself an element of G. We then say that $(G, *)$ is a *group* (often abbreviated to "G is a group" if $*$ may be understood) if it satisfies the following 3 axioms:

I   (associative law) for all g, h, k in G, $(g * h) * k = g * (h * k)$;

II   (existence of two-sided identity) there exists an element e in G such that

$$g * e = e * g = g, \quad \text{for all } g \text{ in } G;$$

III   (existence of two-sided inverses) there exists, to each g in G, an element $\overline{g}$ in G such that

$$g * \overline{g} = \overline{g} * g = e$$

Let us immediately give two examples.

Example 2   Let G be the set of integers (positive, negative, and zero) and let $g * h$ mean $g + h$, the ordinary addition of integers. Then G is a group, with $e = 0$ and $\overline{g} = -g$. Notice that the set of integers does not form a group under multiplication, since there are, in general, no multiplicative inverses, hence the introduction of rational numbers, represented by fractions.

Example 3   Consider the set S of permutations of the set $\{1, 2, 3\}$. There are 6 such permutations; we specify each permutation s by saying where s sends $1, 2, 3$. Thus

$$s_1 : 1 \to 1,\ 2 \to 2,\ 3 \to 3; \quad s_2 : 1 \to 1,\ 2 \to 3,\ 3 \to 2;$$
$$s_3 : 1 \to 2,\ 2 \to 1,\ 3 \to 3; \quad s_4 : 1 \to 2,\ 2 \to 3,\ 3 \to 1;$$
$$s_5 : 1 \to 3,\ 2 \to 1,\ 3 \to 2; \quad s_6 : 1 \to 3,\ 2 \to 2,\ 3 \to 1.$$

The binary operation $s_i * s_j (1 \le i, j \le 6)$ simply produces the permutation s_i *followed by* the permutation s_j. It is obvious that the associative law is satisfied.[5] The identity permutation is s_1. Finally, the inverses are as follows:

The inverse of s_1 is s_1
The inverse of s_2 is s_2
The inverse of s_3 is s_3
The inverse of s_4 is s_5
The inverse of s_5 is s_4
The inverse of s_6 is s_6

It is customary to write, e.g., $\left(\begin{smallmatrix} 1\,2\,3 \\ 2\,1\,3 \end{smallmatrix}\right)$, $\left(\begin{smallmatrix} 1\,2\,3 \\ 3\,1\,2 \end{smallmatrix}\right)$ for the permutations s_3, s_5. There is also a convenient shorthand notation for a *cycle* like $\left(\begin{smallmatrix} 1\,2\,3 \\ 2\,3\,1 \end{smallmatrix}\right)$, in which $1 \to 2, 2 \to 3, 3 \to 1$. (We feel sure you will understand why it's called a cycle.) This shorthand notation is simply (1 2 3); of course, this is the same permutation as (2 3 1) or (3 1 2).

If it is clear from the context that our permutation is operating on the set $\{1, 2, 3\}$, it is convenient to write (1 2) for the permutation s_3, which is a cycle of *length* 2. Of course, very strictly speaking — and sometimes we have to speak strictly — s_3 is the composition of a cycle of length 2 and a cycle of length 1.

---

[5]When the binary operation $g * h$ says, "Do g, then do h," it is always associative. (Do you understand?)

Notice what these two examples have in common, and where they differ. Example 2 is an *infinite* group, and the group operation is *commutative*, that is,

$$g * h = h * g, \quad \text{for all } g, h \text{ in } G.$$

Example 3 is a *finite* group, and the group operation is not commutative, thus

$$s_2 * s_3 = s_4, \quad s_3 * s_2 = s_5$$

In both cases, the group has only one identity element, and each element has only one inverse; moreover, if $\overline{g}$ is the inverse of g, then g is the inverse of $\overline{g}$.

● ● ● **BREAK 2**

(1) Prove that in both of the above examples
   (a) There is only one identity element.
   (b) Each element has only one inverse.
   (c) If $\overline{g}$ is the inverse of g, then g is the inverse of $\overline{g}$.
(2) Try to prove (a), (b), (c) for *any* group G.

Where the group is finite, we can display the entire group law by a square *group table*. Thus, for our second example, the group table is

$*$	s_1	s_2	s_3	s_4	s_5	s_6
s_1	s_1	s_2	s_3	s_4	s_5	s_6
s_2	s_2	s_1	s_4	s_3	s_6	s_5
s_3	s_3	s_5	s_1	s_6	s_2	s_4
s_4	s_4	s_6	s_2	s_5	s_1	s_3
s_5	s_5	s_3	s_6	s_1	s_4	s_2
s_6	s_6	s_4	s_5	s_2	s_3	s_1

(We read off $s_i * s_j$ by looking in row i and column j.) Notice that each row and each column is a permutation of the list of group elements. This is so for any finite group (sometimes this information can be used to complete a table without working out all the individual cases). The *order* of a group is the number of elements in the group, so the order is infinite (Example 2) or some finite number (in the case of Example 3 the order is 6).

Just so that you will feel reassured that we have not forgotten our purpose in introducing you to the idea of a group, let us tell you that the group of Example 3 is called the **symmetric group** on 3 symbols, and written S_3. It will certainly reappear — for example, you will later recognize it as the group of symmetries of an equilateral triangle.

But before we turn to geometry, we require one more result from group theory. This result, due to the great French mathematician Joseph-Louis Lagrange (1736–1813), relates the order of a finite group G and the order of a *subgroup H of G*.

If $(G, *)$ is a group and H is a subset of G, we say that H is a **subgroup** of G if $g_1 * g_2$ belongs to H whenever g_1, g_2 belong to H, and if the induced binary operation on H is a group operation. For this last requirement to be satisfied, it is necessary and sufficient for the identity element to belong to H, and for the inverse of any element in H to belong to H.

Example 2   (revisited) We have the additive group $\mathbb{Z}$ of integers. The subset consisting of even integers is a subgroup, usually written $2\mathbb{Z}$. The subset consisting of non-negative integers is closed under addition, but is *not* a subgroup because the additive inverse of a positive integer is negative.

Before stating Lagrange's Theorem, we introduce some important standard notation. It is customary to write a group G *multiplicatively*, especially if it is not commutative. That is, we write the group operation as multiplication and even (as in ordinary algebra) suppress the product symbol, so that $g_1 * g_2$ is simply written as $g_1 g_2$. Sometimes we may even go further and write the identity element as 1 instead of e; but we will not adopt this notation unless we judge there is no risk at all of confusion. We will, however, always write g^{-1} for the inverse of g. Notice that, since the associative law holds, the positive powers of g, that is, $g, g^2, g^3, \cdots$, have an unambiguous meaning, and so, indeed, do the negative powers if we define

$$g^{-n} = (g^n)^{-1}, \quad \text{for } n \text{ positive} \tag{1}$$

Of course, (1) then also holds for n negative (just as in ordinary algebra).

Notice, too, that the inverse satisfies the basic law

$$(gh)^{-1} = h^{-1}g^{-1}, \quad g, h \text{ in } G \tag{2}$$

Here in (2) we have been careful to state the law so that we do *not* require commutativity.

Now let G be a finite group of order n and let H be a subgroup of G. Of course, H is also a finite group; let its order be m. Then Lagrange's Theorem is this:

Theorem 4 *The order of H divides the order of G, that is, $m|n$.*

Proof[6]   Let g be an element of G (we write $g \in G$), and let gH be the set of all elements gh, as h ranges over the elements of H. We call the set gH a (left) *coset* of H in G, and g a **representative** of the coset gH.

We now prove that, if g_1H, g_2H are two (left) cosets, then either they are disjoint or they coincide. For if $g \in g_1H$, then $g = g_1h_0$ for some $h_0 \in H$, so, for all $h \in H$,

$$gh = g_1h_0h, \quad \text{with} \quad h_0h \in H$$

so $gh \in g_1H$, and thus $gH \subseteq g_1H$. But $g_1 = gh_0^{-1}$, with $h_0^{-1} \in H$, so we may repeat the argument with the roles of g and g_1 exchanged (and h_0^{-1} replacing h_0), concluding that $g_1H \subseteq gH$, so that, finally,[7] $g_1H = gH$. Thus if $g \in g_1H \cap g_2H$, then $g_1H = gH = g_2H$, establishing that g_1H and g_2H coincide if they are not disjoint.

We have next to prove that every coset gH has exactly the same number of elements as H itself. For consider the function φ from H to gH that sends h to gh. This function certainly maps H *onto* gH; for the elements of gH are exactly the elements gh, as h ranges over H. Moreover, φ is one-to-one; for if $gh_1 = gh_2$, then $h_1 = g^{-1}(gh_1) = g^{-1}(gh_2) = h_2$. Thus φ sets up a one-to-one correspondence[8] between H and gH.

Now suppose that H has m elements. Then every (left) coset has m elements, and G, as a set, is the disjoint union of a certain number of cosets. If, then, G is the disjoint union of k cosets, we have proved that

$$n = km \qquad (3)$$

establishing Lagrange's Theorem.   □

---

[6]We advise you to read this argument carefully. It is really subtle.

[7]We have given you here an argument that does *not* depend on G or H being finite.

[8]See previous footnote.

Remarks

(i) The number k that appears in (3) is called the **index** of H in G.

(ii) Notice that we could have argued with right cosets Hg. Obviously, we would have arrived at the same relationship (3), so the index is also the number of disjoint right cosets. This is significant because it is by no means true in general that a left coset is a right coset.

● ● ● **BREAK 3**

(1) Let G be the additive group of remainders modulo 12 (think of a 12-hour clock). What is the order of G? Find all the subgroups of G and list their orders.

(2) Set up a one-to-one correspondence between the set of left cosets of H in G and the set of right cosets of H in G. (Be careful — there is no well-defined function sending the coset gH to the coset Hg.)

This is as far as we need to take the group theory in order to be precise as to what we mean by a *geometry* on a configuration A, and the symmetry of the resulting geometric figure. However, some further (and obviously relevant) group theory will be found in Section 5.

We have been guided in our definitions by the approach of the German mathematician Felix Klein (1849–1925) to explaining the nature of geometry. Consider, for example, the usual plane Euclidean geometry, in which we study the properties of planar figures that are invariant under certain *Euclidean motions*. These motions certainly include *translation* and *rotation* in the plane of the figure, but it is a matter of choice whether they include *reflection*. For example, the elaborate 7-gon[9] of Figure 1(a) is invariant under rotations through $\frac{2\pi}{7}$ about its center, but, unlike the ordinary regular 7-gon, not under reflection about any axis through its center. Thus, to define our geometry, we must decide whether we allow Euclidean motions that reverse orientation. Of course, if we allow certain Euclidean motions, we must also allow compositions and inverses of such motions, so we postulate a certain *group* G of allowed motions of the *ambient* plane, the plane containing our planar figure. If A is such a planar figure, then,

---

[9]Those of you who are interested in knowing how to construct this figure (by simply folding paper) may wish to consult Chapter 4.

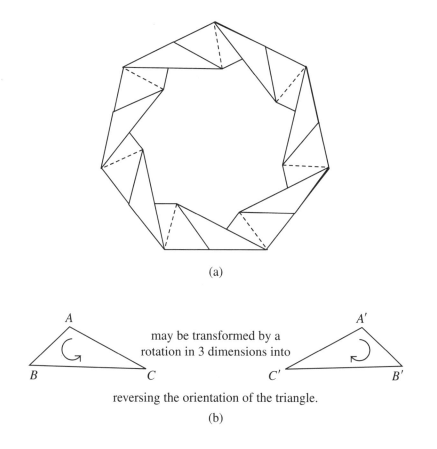

(a)

A   may be transformed by a   *A′*

rotation in 3 dimensions into

B   *C*   *C′*   *B′*

reversing the orientation of the triangle.

(b)

FIGURE 1

for any $g \in G$, Ag is again a planar figure.[10] In *the G-geometry of A*, we study the properties of the figure A that it shares with all the figures Ag as g varies over G. Such properties are called the **G-invariants** of A, abbreviated to **invariants** if, but only if, the group G may be understood. It is thus the group G that, according to Klein, determines the geometrical nature of A.

Example 5 Let G be the group of motions of the plane generated by translations, rotations, and reflections (in a line); we call this the ***Euclidean***

---

[10]We prefer to write Ag for the image of A under the motion g, rather than gA. This is because, in the group G, gh means "first g, then h."

group in 2 dimensions and write it E_2. Then the Euclidean geometry of the plane is the study of the properties of subsets of the plane that are invariant under the motions in E_2. For example, the property of being a polygon is a Euclidean property; the number of vertices and sides of a polygon is a Euclidean invariant. On the other hand, as we have hinted, orientation is not invariant with respect to this group, though it would be if we disallowed reflections. Thus, by means of a motion in E_2 the triangle ABC may be turned over (flipped) to form the triangle $A'B'C'$ as shown in Figure 1(b). But the orientation $\overrightarrow{ABC}$ is counterclockwise, while the orientation $\overrightarrow{A'B'C'}$ is clockwise.

Example 6 We may step up a dimension, passing to the group E_3 of Euclidean motions in 3-dimensional space. Notice that it is natural to think of reflections in a line (of a planar figure) as a motion in space, since it can be achieved by a rotation in some suitable ambient 3-dimensional space containing the plane figure — you will surely see this by looking again at Figure 1(b). However, it requires a greater intellectual effort to think of reflection in a plane (of a spatial figure) as a motion in some ambient 4-dimensional space! Who would think of turning the golden dodecahedron[11] (see Figure 2(c)) inside out? Thus it is common not to include such reflections in defining 3-dimensional geometry. This preference is, of course, a consequence of our experience of living in a 3-dimensional world and has no mathematical basis. However, whenever we are highlighting the construction of actual physical models of geometrical configurations, it is entirely reasonable to omit motions to which the models themselves cannot be subjected.

We now move to a precise definition of symmetry. Let a geometry be defined on the ambient space of a configuration A by means of the group of motions G. Then the *symmetry group* of A, relative to the geometry defined by G, is the subgroup G_A of G consisting of those motions $g \in G$ such that $Ag = A$, that is, those motions that map A onto itself, or, as we say, under which A is *invariant*. Thus, for example, if our geometry is defined by rotations and translations in the plane, and if A is an equilateral triangle, then its symmetry group G_A consists of rotations about its center through $0°$, $120°$, and $240°$; if, in our geometry, we also allow reflections, then the symmetry group has 6 elements instead of 3, and is, in fact, the

---

[11]We have included Figures 2(a), 2(b) as well for the benefit of those readers who might wish to construct the golden dodecahedron. See Figure 33 of Chapter 4.

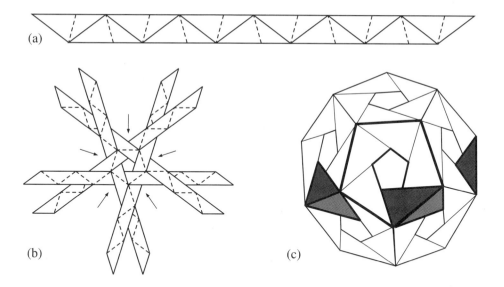

FIGURE 2 The golden dodecahedron. This extraordinarily beautiful model may be made by braiding together six colored strips like the one shown in Figure 2(a). Begin by laying out five of the strips as shown in Figure 2(b) and then attach paper clips at the places marked by the arrows. Next continue to braid these strips together, remembering what the finished model will look like (attaching paper clips, in the middle of the faces—not at the vertices—wherever needed to hold everything in position). At a certain stage you will need to insert the sixth strip about the "equator." When you are finished, every strip should go over and under the strips it meets all the way around the model, and there should be no loose ends sticking out. For more detailed information about how to prepare the strips, consult Chapter 4.

very well-known group S_3 (recall Example 3), called the ***symmetric group on 3 symbols*** — the symbols may be thought of in this case as the vertices of the triangle. We repeat, for emphasis, that the symmetry group G_A of the configuration A is a *relative* notion, depending on the choice of geometry G.

It is plain that no compact (bounded) configuration can possibly be invariant under a non-zero translation. Thus when we are considering the symmetry group of such a figure we may suppose G to be generated by rotations and, perhaps, reflections. Moreover, any such motion in the plane is determined by its effect on 3 independent points, and any such motion in 3-dimensional space is determined by its effect on 4 independent points. Since a (plane) polygon has at least 3 vertices and a polyhedron has at least 4 vertices, and since any element of the symmetry group of a polygon or a polyhedron must map vertices to vertices, it follows that the symmetry group of a polygon or a polyhedron is *finite* (compare the symmetry groups of a circle or a sphere).

The symmetry group of any polygon with n sides is, by the argument above, a subgroup of S_n, the group of permutations of n symbols, also called the **symmetric group** on n symbols. If G is generated by rotations alone, and the polygon is regular, its symmetry group is the cyclic group[12] of order n, often written C_n, generated by a rotation through an angle of $\frac{2\pi}{n}$ radians about the center of the polygonal region. If G also includes reflections, then the symmetry group has $2n$ elements and includes n reflections; this group is called the **dihedral group** of order $2n$ and is usually written D_n.

In discussing the symmetry groups of *polyhedra*, however, we will, as indicated earlier, always assume that the geometry is given by the group G generated by rotations in 3-dimensional space. Then the symmetry group of the regular tetrahedron is the so-called **alternating group** A_4. In general, A_n is the subgroup of S_n consisting of the *even* permutations[13] of n symbols; it is of index 2 in S_n, whose order is $n!$, so that its order is $\frac{1}{2}(n!)$. Thus the order of A_4 is 12.

The cube and the regular octahedron have the same symmetry group, namely S_4. It is easy to see why the symmetry groups are the same; for the centers of the faces of a cube are the vertices of a regular inscribed octahedron, and the centers of the faces of a regular octahedron are the vertices of an inscribed cube. Likewise, and for the same reason, the regular dodecahedron and the regular icosahedron have the same symmetry group, which is A_5. It is a matter of interest and relevance here that the elements of the symmetry group of the diagonal cube (Figure 3, which is Figure 27 of Chapter 4) permute the four braided strips from which the model is made, and thus the symmetry group is S_4. (We'll have you study this closer in Break 6.)

We are now in a position to give at least one possible precise meaning to the statement "Figure A is more symmetric than Figure B." We assume that A and B are defined as geometric figures by the same group of motions G. If it happens that the symmetry group G_A of A strictly contains the symmetry group G_B of B, then we are surely entitled to say that A is "more symmetric" than B. Notice that the situation described may, in fact, occur because B is obtained from A by adding features that destroy some

---

[12] A cyclic group of order n is a group of order n, all of whose elements are powers of a given element g, called a **generator** of the group.

[13] See Section 5 of this chapter for a careful treatment of even and odd permutations.

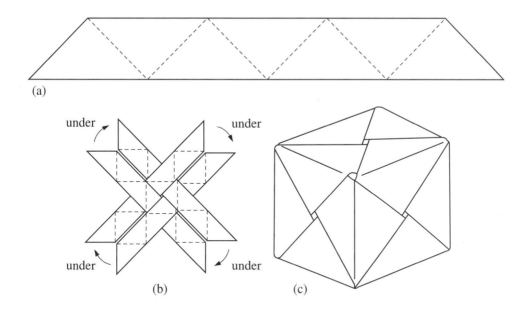

FIGURE 3 To construct this model, begin with four strips, each one consisting of 7 isosceles right triangles as shown in Figure 3(a). Each strip should be of a different color. Then fold the strips so that there are "mountain" folds on the colored side of the paper. With the mountain folds down, lay the strips out as shown in Figure 3(b). Secure the center of the configuration with a small piece of tape and then continue to "braid" the strips together to form the diagonal cube shown in Figure 3(c). Every face should have four colors visible, and every strip should go over and under the strips it meets as it goes around the cube. Finally, all the ends should be tucked in.

of the symmetry of A. For example, Figure 4(a) shows a typical strip that may be used to braid each of the Platonic solids. (Precise details about the construction of these models appear in Chapter 4.) However, the resulting braided Platonic solids of Figure 4(b) will all have their symmetry reduced if we braid them from strips of different colors (or different patterns). It is natural to ask whether it is possible to braid the tetrahedron, octahedron, and icosahedron in such a way as to retain all the symmetry of the original polyhedron.[14] We will address this question in the next break.

Meanwhile, returning to the discussion of symmetry, we note that the notion "more symmetric" above is really too restrictive. For we would like to be able to say that the regular n-gon becomes more symmetric as n increases. We are thus led to a weaker notion that will be useful if we

[14]Notice that we already know how to do this for the cube (Figure 3) and the dodecahedron (Figure 2).

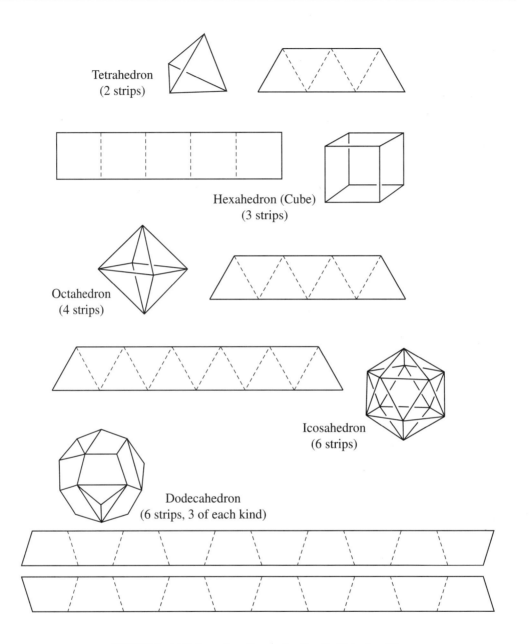

FIGURE 4   **(a) Pattern pieces for constructing the Platonic solids.**

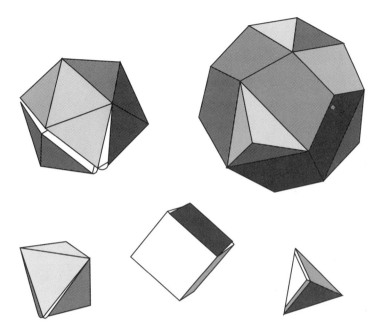

FIGURE 4   (b) The completed Platonic solids.

are dealing with figures with finite symmetry groups (e.g., polygons and polyhedra). We could then say — and do say — that A is more symmetric than B if G_A has more elements than G_B. Thus we have, in fact, two notions whereby we may compare symmetry — and they have the merit of being consistent. Indeed, if A is more symmetric than B in the first sense, it is more symmetric than B in the second sense — but not conversely.

Notice that we deliberately avoid the statement — often to be found in popular writing — "A is a symmetric figure." We regard this statement as having no precise meaning!

● ● ● BREAK 4

(1) Why should we wish to say that the regular n-gon becomes more symmetric as n increases? What happens to the regular n-gon as n increases indefinitely?

(2) Figure 5(b) shows a typical straight strip of 5 equilateral triangles with a slit in each triangle from the top (or bottom) edge to (just

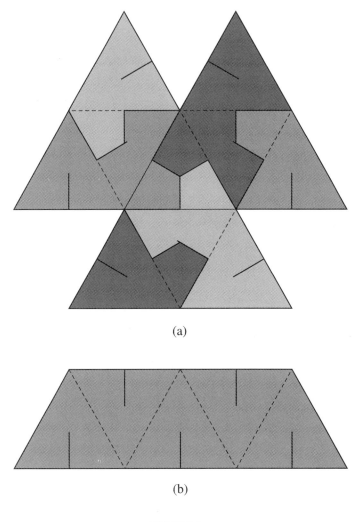

(a)

(b)

FIGURE 5

past) the center.[15] A tetrahedron with symmetry group A_4 may be constructed out of three of these strips. Figure 5(a) shows how the three strips are interlaced initially. We leave the completion of the model as a challenge to you.

[15]Theoretically, the slit could go just to the center, but the model is then impossible to assemble. You need to have some leeway for the pieces to be free to move during the process of construction — although they will finally land in a symmetric position, so that it *looks* as though the slit need not have gone past the center.

(3) **Harder.** Figure 6(a) shows the layout of three strips for the beginning of the construction of an octahedron with symmetry group S_4. Can you build it? We'll give you one more hint. When you use the layout of Figure 6(a) remember that the strip shown below it in Figure 6(b) has to be braided into the figure above it.

(4) **Much harder.** An icosahedron with symmetry group A_5 may be constructed from 6 strips of the type shown in Figures 5 and 6, where each strip has 11 equilateral triangles. Over to you! But

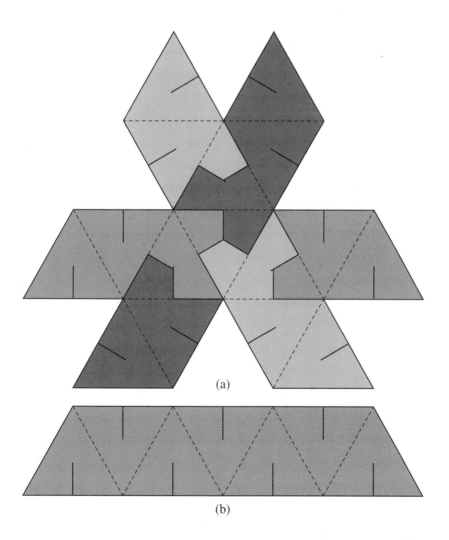

(a)

(b)

FIGURE 6

FIGURE 6   (c)

take heart — these models take several hours to construct. Just to prove that they really do exist we show a photo of them in Figure 6(c).

8.3 HOMOLOGUES

George Pólya, who made great contributions not only to mathematics itself, but also to the understanding of how and why we do mathematics — or perhaps one should say, "how and why we should do mathematics" — was particularly fascinated by the Platonic solids and first introduced his notion of *homologues* (orally) in connection with the study of their symmetry. We know that the idea of homologues played a key role in his thinking about one of his greatest contributions to the branch of mathematics known as *combinatorics*, namely, the ***Pólya Enumeration Theorem*** (see [5] for an intuitive account and [3] for a detailed account). Let us describe this notion of homologues in terms of symmetry groups. We believe that we are thereby increasing the scope of the notion while entirely maintaining the spirit of Pólya's original idea.[16]

Let A be a geometrical configuration in a geometry given by a group Γ, and let the symmetry group of A with respect to this geometry be Γ_A; and

[16]We will ourselves be discussing the Pólya Enumeration Theorem in the next section.

let B be a subset of A. Thus, for example, A may be a polyhedron and B a face of that polyhedron. We consider the subgroup Γ_{AB} of Γ_A consisting of those motions in the symmetry group Γ_A of A that map B to itself. We consider a right coset of Γ_{AB} in Γ_A, that is, a set $\Gamma_{AB}g$, $g \in \Gamma_A$. (Be sure you understand what Γ_{AB} and $\Gamma_{AB}g$ mean before you read on.) Every element in $\Gamma_{AB}g$ sends B to the same subset Bg of A. The collection of these subsets is what Pólya called the collection of **homologues** of B in A. We see that the set of homologues of B in A is in one-to-one correspondence with the set of right cosets of Γ_{AB} in Γ_A.

Example 7 Consider the pentagonal dipyramid A of Figure 7(b). We may specify any motion in the symmetry group of A by the resulting permutation of its vertices $1, 2, 3, 4, 5, 6, 7$; note that the polar vertices are 6 and 7. In fact, Γ_A is the dihedral group D_5, with 10 elements, given by the following permutations:

$$
(1\,2\,3\,4\,5\,6\,7) \begin{cases}
\to (1\,2\,3\,4\,5\,6\,7) & \text{(identity)} \\
\to (2\,3\,4\,5\,1\,6\,7) & \text{(rotation through } \tfrac{2\pi}{5} \text{ about axis 67)} \\
\to (3\,4\,5\,1\,2\,6\,7) & \text{(rotation through } \tfrac{4\pi}{5} \text{ about axis 67)} \\
\to (4\,5\,1\,2\,3\,6\,7) & \bullet \\
\to (5\,1\,2\,3\,4\,6\,7) & \bullet \\
\to (5\,4\,3\,2\,1\,7\,6) & \text{(interchanging the poles)} \\
\to (4\,3\,2\,1\,5\,7\,6) & \text{(interchange plus rotation} \\
 & \quad \text{through } \tfrac{2\pi}{5}) \\
\to (3\,2\,1\,5\,4\,7\,6) & \text{(interchange plus rotation} \\
 & \quad \text{through } \tfrac{4\pi}{5}) \\
\to (2\,1\,5\,4\,3\,7\,6) & \bullet \\
\to (1\,5\,4\,3\,2\,7\,6) & \bullet
\end{cases}
$$

First, let B be the edge 16. Then $\Gamma_{AB} = \{\text{Id}\}$, since only the identity sends the subset $(1, 6)$ to itself. Thus the index of Γ_{AB} in Γ_A is ten, and there are ten homologues of the edge 16; these are the ten "spines" of the dipyramid (i.e., we exclude the edges around the equator). Second, let B be the edge 12. Then Γ_{AB} has two elements, since there are two elements of Γ_A, namely the identity and $(1\,2\,3\,4\,5\,6\,7) \to (2\,1\,5\,4\,3\,7\,6)$, which send the subset $\{1, 2\}$ to itself. Thus the index of Γ_{AB} in Γ_A is five, and there are five homologues of the edge 12; these are the five edges around the equator.

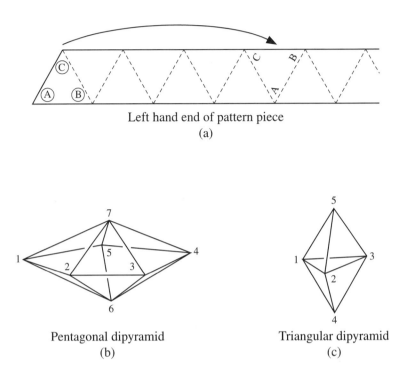

Left hand end of pattern piece
(a)

Pentagonal dipyramid
(b)

Triangular dipyramid
(c)

FIGURE 7 **The pentagonal dipyramid may be made from a strip of 31 equilateral triangles. You begin the constructions by placing the first triangle over the eighth triangle as shown in Figure 7(a). At that point just let the rest of the strip fall around the model, remembering what it should look like when you are finished. The last triangle will tuck into a slot produced by an edge of the tape from a previous crossing of the last face. A similar construction of the triangular dipyramid exists, beginning with a strip of 19 equilateral triangles. More detailed instructions for preparing the strips may be found in Chapter 4.**

Third, let B be the face 126. Then $\Gamma_{AB} = \{\mathrm{Id}\}$, so that, as in the first case, there are 10 homologues of the face 126; in other words, all the (triangular) faces are homologues of each other.

● ● ● **BREAK 5**

(1) Consider the triangular dipyramid P of Figure 7(b). Specify the motions of the symmetry group by writing down the resulting permutations of its vertices 1, 2, 3, 4, 5; note that the polar vertices are 4 and 5. In fact, Γ_P is the dihedral group D_3, with 6 elements.

(2) How many homologues does the edge 12 have? How many homologues does the vertex 4 have? How many homologues does the face 124 have?

Let us now explain the Pólya Enumeration Theorem — actually, there are *two* theorems — and see how the notion of homologue fits into the story.

8.4   THE PÓLYA ENUMERATION THEOREM

Let S be a finite set; the reader might like to keep in mind the set of vertices (or edges, or faces) of a polygon or polyhedron. Let G be a finite (symmetry) group acting on S. Suppose S has n elements, and that G has m elements; we write $|S| = n$, $|G| = m$. We may represent the elements of the set S by the integers $1, 2, \cdots, n$. If $g \in G$, then g acts as a *permutation* of $\{1, 2, \cdots, n\}$. Now every permutation is uniquely expressible as a composition of cyclic permutations on mutually exclusive subsets of the elements of S. For example, the permutation

$$\begin{pmatrix} 1\ 2\ 3\ 4\ 5\ \ 6\ \ 7\ 8\ 9\ 10\ 11 \\ 2\ 4\ 7\ 1\ 3\ 11\ 5\ 6\ 8\ 10\ \ 9 \end{pmatrix} \tag{4}$$

of the set S of integers $\{1, 2, \cdots, 11\}$ is the composite

$$(1\ 2\ 4)(3\ 7\ 5)(6\ 11\ 9\ 8)(10)$$

where, e.g., $(1\ 2\ 4)$ denotes the cyclic permutation

$$\begin{pmatrix} 1\ 2\ 4 \\ 2\ 4\ 1 \end{pmatrix}$$

Thus the permutation (4) is the composite of one cyclic permutation of length 1, two cyclic permutations of length 3, and one cyclic permutation of length 4, the cyclic permutations acting on disjoint subsets of the set S. In general, a permutation of S has the *type* $\{a_1, a_2, \cdots, a_n\}$ if it consists of a_1 permutations of length 1, a_2 permutations of length 2, $\cdots$, a_n permutations of length n, the permutations having disjoint domains of action; notice that $\sum_{i=1}^{n} i a_i = n$. For example, the permutation (4) has type $\{1, 0, 2, 1, 0, 0, 0, 0, 0, 0, 0\}$. If g has type $\{a_1, a_2, \cdots, a_n\}$, we define the *cycle index* of g to be the monomial

$$Z(g) = Z(g; x_1, x_2, \cdots, x_n) = x_1^{a_1} x_2^{a_2} \cdots x_n^{a_n}$$

The **cycle index of G** is $Z(G) = Z(G; x_1, x_2, \cdots, x_n) = \frac{1}{m} \sum_{g \in G} Z(g)$.

FIGURE 8   We denote this labeling of the vertices of the square by $\frac{1}{4}\blacksquare\frac{2}{3}$.

We give an example that we will revisit periodically throughout this section.

Example 8   We consider the symmetries[17] of the square that is shown in Figure 8.

The group G of symmetries is the group D_4 of order 8, which we here describe as a group of permutations of the set S of vertices $\{1, 2, 3, 4\}$. Thus $n = 4$, $m = 8$, and the elements of G are

g_1 (identity)	$(1)(2)(3)(4)$	cycle index	x_1^4
g_2	$(1\ 2\ 3\ 4)$	cycle index	x_4
g_3	$(1\ 3)(2\ 4)$	cycle index	x_2^2
g_4	$(1\ 4\ 3\ 2)$	cycle index	x_4
g_5	$(1\ 3)(2)(4)$	cycle index	$x_1^2 x_2$
g_6	$(2\ 4)(1)(3)$	cycle index	$x_1^2 x_2$
g_7	$(1\ 2)(3\ 4)$	cycle index	x_2^2
g_8	$(1\ 4)(2\ 3)$	cycle index	x_2^2

Thus the cycle index of G, obtained from the calculations above, is

$$\frac{1}{8}\left(x_1^4 + 2x_1^2 x_2 + 3x_2^2 + 2x_4\right)$$

We leave the example temporarily and return to the general case. Suppose we want to *color* the elements of S. That is, we have a finite set Y of colors, $|Y| = r$, and then a *coloring* of S is a function[18] $f : S \to Y$. For any $g \in G$, we regard the colorings f and fg as indistinguishable or

[17]Recall that, in considering the symmetries of a polygon, we permit reflections in three-dimensional space about a line.

[18]We speak of a *coloring* of S; this may be literally true, or it may merely be a metaphor for a rule for dividing the elements of S into disjoint classes.

equivalent; and a *pattern* is an equivalence class of colorings. Then Pólya's first theorem is as follows.

Theorem 9 *For a given group of symmetries G and a given set of colors Y with $|Y| = r$, the number of patterns is $Z(G; r, r, \cdots, r)$.*

Example 8 (continued) Suppose the vertices of the square are to be colored using r colors. Then the number of patterns is $\frac{1}{8}(r^4 + 2r^3 + 3r^2 + 2r)$. If, in particular, the vertices are to be colored red or blue, then $r = 2$, and the number of patterns is $\frac{1}{8}(16 + 16 + 12 + 4) = 6$. In fact, the patterns are represented by the 6 colorings

$$
\begin{matrix} R & R \\ & \blacksquare \\ R & R \end{matrix} \quad
\begin{matrix} B & R \\ & \blacksquare \\ R & R \end{matrix} \quad
\begin{matrix} B & B \\ & \blacksquare \\ R & R \end{matrix} \quad
\begin{matrix} B & R \\ & \blacksquare \\ R & B \end{matrix} \quad
\begin{matrix} R & B \\ & \blacksquare \\ B & B \end{matrix} \quad
\begin{matrix} B & B \\ & \blacksquare \\ B & B \end{matrix}
$$

Notice that we regard the colorings $\begin{smallmatrix}B&&R\\&\blacksquare&\\R&&R\end{smallmatrix}$ and $\begin{smallmatrix}R&&B\\&\blacksquare&\\R&&R\end{smallmatrix}$, for example, as indistinguishable or equivalent; we are sure you see why.

We now describe Pólya's second theorem. This is really the big theorem, and the first theorem is, in fact, deducible from it. Let us enumerate the elements of Y (the colors) as $y_1, y_2, \cdots, y_r$.

Theorem 10 (**The Pólya Enumeration Theorem**) *Evaluate the cycle index*

$$
Z(G; x_1, x_2, \ldots, x_n) \quad at \quad x_i = \sum_{j=1}^{r} y_j^i, \quad i = 1, 2, \ldots, n.
$$

Then the coefficient of $y_1^{n_1} y_2^{n_2} \cdots y_r^{n_r}$ is the number of patterns assigning the color y_j to n_j elements[19] of S.

Example 8 (continued) For the symmetries of the square we know that

$$
Z(G) = \frac{1}{8}(x_1^4 + 2x_1^2 x_2 + 3x_2^2 + 2x_4)
$$

Let $Y = \{R, B\}$; then the evaluation of $Z(G)$ at $x_i = R^i + B^i, i = 1, 2, 3, 4$, yields

$$
\frac{1}{8} \left((R + B)^4 + 2(R + B)^2(R^2 + B^2) + 3(R^2 + B^2)^2 + 2(R^4 + B^4) \right)
$$

$$
= R^4 + R^3 B + 2R^2 B^2 + R B^3 + B^4
$$

---

[19] Of course, $\sum_{j=1}^{r} n_j = n$.

(It is, of course, no coincidence that this polynomial is homogeneous, of degree $|S|$, and symmetric; and has integer coefficients.)

Thus the Pólya Enumeration Theorem tells us that there is one pattern with 4 red vertices (obvious); one pattern with 3 red vertices and 1 blue vertex (represented by the coloring $^B_R\blacksquare^R_R$); 2 patterns with 2 red vertices and 2 blue vertices (represented by the colorings $^B_R\blacksquare^B_R$ and $^B_R\blacksquare^R_B$); and the remaining possibilities are most easily analyzed by considering the symmetric roles of R and B.

Consider the various coloring functions $S \to Y$ representing a given pattern. These functions all have the form $fg : S \to Y$, where f is a fixed coloring and g ranges over the elements of G. It will turn out that the colorings fg are essentially the *homologues of f*. Let us first revert to our example.

Example 8　(continued) As we have seen, there is one coloring in which all vertices of the square are colored red. There is only one homologue, namely, $^R_R\blacksquare^R_R$.

There is one coloring in which 3 vertices are colored red and one blue. There are 4 homologues, namely,

$$^B_R\blacksquare^R_R \qquad ^R_R\blacksquare^B_R \qquad ^R_R\blacksquare^R_B \qquad ^R_B\blacksquare^R_R$$

There are two colorings in which 2 vertices are colored red and 2 blue. In the first there are 4 homologues, namely,

$$^B_R\blacksquare^B_R \qquad ^R_R\blacksquare^B_B \qquad ^R_B\blacksquare^R_B \qquad ^B_B\blacksquare^R_R$$

In the second there are 2 homologues, namely,

$$^B_R\blacksquare^R_B \qquad ^R_B\blacksquare^B_R$$

You may complete the analysis by considerations of symmetry.

Let us now show how this concept of homologues agrees with our earlier definition. We are given the group G of permutations[20] of S. Given a coloring $f : S \to Y$, we consider the subset G_0 of G consisting of those g such that $fg = f$, that is, those movements of S that preserve the coloring. It is easy to see (just as easy as in our earlier, simpler situation) that G_0 is a *subgroup* of G. Corresponding to each coset $G_0 g$ of G_0 in G we have a coloring fg of S, and these colorings run through the pattern determined by f. We describe the set of colorings $\{fg\}$ in this pattern as the set of *homologues* of the coloring f; just as in our geometric definition

---

[20]Recall, from Section 3, the group Γ_A of symmetries of the configuration A.

in Section 3, they are in one-to-one correspondence with the set of right cosets of G_0 in G.

● ● ● **BREAK 6**

(1) Construct the diagonal cube[21] shown in Figure 3.

(2) Use this cube to fill in the following table:

Description of rotation about the axis through the centers of	The amount of rotation	The number of rotations of this type
Opposite faces	$\pm\frac{1}{4}$ turn	6
Opposite faces	$\frac{1}{2}$ turn	
Opposite vertices	$\pm\frac{1}{3}$ turn	
Opposite edges	$\frac{1}{2}$ turn	
Identity	0	
		Total number of rotations $=$

(3) Fill in the following table about the permutations of four objects, and make the obvious comparison of the two tables.

Description of the type of permutation	The number of permutations of this type
$(\_\_\_\_) \left(\begin{smallmatrix}\text{one cycle of}\\ \text{length 4}\end{smallmatrix}\right)$	6
$(\_\_)(\_\_) \left(\begin{smallmatrix}\text{two cycles, each}\\ \text{of length 2}\end{smallmatrix}\right)$	
$(\_\_\_)(\_) \left(\begin{smallmatrix}\text{one cycle of length 3}\\ \text{and one cycle of length 1}\end{smallmatrix}\right)$	
$(\_\_)(\_)(\_) \left(\begin{smallmatrix}\text{one cycle of length 2}\\ \text{and two cycles of length 1}\end{smallmatrix}\right)$	
$(\_)(\_)(\_)(\_) \left(\begin{smallmatrix}\text{four cycles of length 1,}\\ \text{i.e., the identity}\end{smallmatrix}\right)$	
	Total number of $=$ permutations

---

[21]Notice that the diagonal cube is so-called because of the appearance of the diagonals on each of its faces. We are not referring to the *interior* diagonals of the cube.

(4) Notice anything? (Hint: Think of the strips of your cube as being numbered 1, 2, 3, 4 and observe what happens to the strips when you perform the rotations in part (2). This is one very vivid way to see why the symmetry group of the cube is S_4. Of course, there are other ways, too. For example, the interior diagonals of the cube might be labeled 1, 2, 3, 4, and you would see that the rotations of the cube simply permute these diagonals. It is just a little harder to "see" the interior diagonals than it is to see the strips of the diagonal cube.

(5) Write the cycle index for the 4 strips of the diagonal cube.

(6) Use the Pólya Enumeration Theorem to determine how many different diagonal cubes can be made with four strips if you may choose any one of four colors for each strip.

(7) Write the cycle index for the 6 strips of the Golden Dodecahedron.

(8) Use the Pólya Enumeration Theorem to determine how many different Golden Dodecahedra can be made with six strips, each a different color.

★ We are now prepared to offer our general definition of a homologue. To help you to follow this, we first repeat briefly the two contexts in which this notion has already arisen in this chapter.

First, consider a geometric configuration Z with respect to a geometry given by the group Γ. Let G be the subgroup of Γ consisting of those $g \in \Gamma$ such that $Zg = Z$. So G is the symmetry group of Z in the geometry given by Γ; G acts on Z.

$$\text{Set} \quad X = 2^Z = \text{set of subsets of } Z.$$

Now G acts on X. Then $x \in X$ is a subset of Z, and G_x is the subgroup of G consisting of those g that map x to itself.[22] Hence we obtain the set of homologues of x; and the homologues are in one-to-one correspondence with the cosets of G_x in G.

Second, recall that, in this section, we have discussed the set of functions $f : S \to Y$, or colorings of S, and a group G acting on S.

$$\text{Let} \quad X = Y^S = \text{set of colorings of } S, \text{ i.e., functions from } S \text{ to } Y$$

Then G acts on X by the rule

$$fg(s) = f(sg), \quad g \in G, \quad f : S \to Y, \quad s \in S$$

---

[22]In Section 3 we had $Z = A$, $G = \Gamma_A$, $x = B$, $G_x = \Gamma_{AB}$.

Now, for any f, $G_f = \{g \,|\, fg = f\}$, and we get a one-to-one correspondence between the homologues of f, as described in this section, and the cosets of G_f in G. The homologues of f are the colorings of S determining the same pattern as f.

Thus the common description, covering both applications, is this: A group G acts on a set S. Let Y be a fixed set and consider the induced action of G on Y^S. If $f \in Y^S$, that is, if f is a function from S to Y, let G_f be the subgroup of G consisting of those $g \in G$ that fix f, that is, such that $fg = f$. Then the ***homologues*** of f (with respect to G) are the functions fg, as g ranges over G; and they are in one-to-one correspondence with the cosets of G_f in G, under the rule $G_f g \longleftrightarrow fg$. In our first example above, we had $S = Z$, $Y = \{0, 1\}$ (so that Y^S is the set of subsets of Z), and $f = x$.

This fulfills our promise to George Pólya to write up his idea of ***homologues***.

We will give you a further example of homologues in Break 10; but before we can make that example clear we need to describe *even* and *odd* permutations and the *alternating group*.

8.5 EVEN AND ODD PERMUTATIONS

We described the symmetry group of a regular tetrahedron as the alternating group A_4 consisting of *even* permutations of the set $(1, 2, 3, 4)$. We owe you at the very least a definition of this concept! The reason we postponed giving you this is that the *precise* definition is difficult to understand. See if you agree with us.

Let A be the *alternating form in n variables*, that is, the expression

$$A = \prod_{1 \leq i < j \leq n} (x_i - x_j) \tag{5}$$

Then any permutation ρ in S_n acts on A by permuting the suffixes $i, j, \cdots$ appearing in (5) by means of ρ; thus

$$A\rho = \prod_{1 \leq i < j \leq n} (x_{i\rho} - x_{j\rho}) \tag{6}$$

We claim that $A\rho$ is either A or $-A$. For, given any factor $x_i - x_j$ of $A\rho$, then $i = h\rho$, $j = k\rho$, for some (unique) h, k between 1 and n, and either $x_h - x_k$ or $x_k - x_h$ occurred in A. Thus we are led to the following crucial definition:

Definition 11 The permutation ρ is *even* if $A\rho = A$; it is *odd* if $A\rho = -A$.

We call the evenness or oddness of a permutation its **parity**.

Example 12 Consider the *cyclic* permutation $(1\ 2\ 3)$, that is, $\rho(1) = 2$, $\rho(2) = 3$, $\rho(3) = 1$. Then if $A = (x_1 - x_2)(x_1 - x_3)(x_2 - x_3)$,

$$A\rho = (x_2 - x_3)(x_2 - x_1)(x_3 - x_1) = A$$

Thus ρ is even. On the other hand, if σ is the permutation $(1\ 3)$, that is, $\sigma(1) = 3$, $\sigma(2) = 2$, $\sigma(3) = 1$, then

$$A\sigma = (x_3 - x_2)(x_3 - x_1)(x_2 - x_1) = -A,$$

so σ is odd.

Now in the next break you will see that every permutation may be expressed as a composition of *transpositions*, that is, of cycles that (like $(1\ 3)$ above) merely interchange two numbers. Granted this, one may show the following:

Theorem 13 *A permutation is even if and only if it may be expressed as a composition of an even number of transpositions.*

You may ask — why don't we simply *define* an even permutation as one that may be expressed as a composition of an even number of transpositions? For that is surely the reason for the use of the name "even." The answer is that it is by no means obvious that, if a permutation can be expressed as a composition of an *even* number of transpositions, it cannot also be expressed as a composition of an *odd* number of transpositions. It is, indeed, Theorem 13 that makes that fact clear.

However, before we prove Theorem 13, we'll take the promised break and put you to work seeing why every permutation may be expressed as a composition of transpositions.[23]

● ● ● BREAK 7

(1) Satisfy yourselves, by looking at a particular but not special case, that every permutation of a set of numbers may be expressed as a composition of cycles acting on disjoint sets of numbers. (See our example (4).)

---

[23]We adopt the usual convention that the identity permutation is the *empty* composition of transpositions.

(2) We now need to show that every cycle $(\iota_1 \iota_2 \cdots \iota_r)$ is a composition of transpositions. Of course, this holds if $r = 2$. Now, if $r \geq 3$, show that

$$(\iota_1 \iota_2 \cdots \iota_r) = (\iota_1 \iota_2 \cdots \iota_{r-1})(\iota_1 \iota_r)$$

(recall that fg means "do f, then do g") and hence deduce, by induction on r, that the cycle $(\iota_1 \iota_2 \cdots \iota_r)$ is a composition of transpositions.

(3) Find an explicit expression for $(\iota_1 \iota_2 \cdots \iota_r)$ as a composition of transpositions.

Now we return to the proof of Theorem 13. It is obvious that even and odd permutations under composition $*$ act like ordinary integers under addition, that is,

$$\text{even} * \text{even} = \text{even},$$
$$\text{odd} * \text{odd} = \text{even},$$
$$\text{even} * \text{odd} = \text{odd},$$
$$\text{odd} * \text{even} = \text{odd}.$$

It therefore follows immediately that, if we can show that every transposition is odd, then it must be true that any permutation expressible as a composition of an *even* number of transpositions is *even*, and any permutation expressible as a composition of an *odd* number of transpositions is *odd* — and this will prove Theorem 13. Thus we are left to prove that *every transposition is odd*.

First, we claim that the transposition $(1\ 2)$ is odd. For if we apply $(1\ 2)$ to the form $A = \prod_{1 \leq i < j \leq n} (x_i - x_j)$, then the factor $(x_1 - x_2)$ is transformed into its negative, while all the other factors are merely permuted among themselves with no change of sign. Thus $(1\ 2)$ is an odd permutation. Second, consider the transposition $(i\ j)$, $i < j$. We claim that

$$(2\ j)(1\ 2)(2\ j) = (1\ j), \quad j \geq 3,$$
$$(1\ j)(1\ 2)(1\ j) = (2\ j), \quad j \geq 3,$$
$$(1\ i)(2\ j)(1\ 2)(1\ i)(2\ j) = (i\ j), \quad i \geq 3, \quad j \geq 3.$$

Thus, since $(1\ 2)$ is odd, so is every $(i\ j)$, and Theorem 13 is completely proved.[24]

---

[24]Notice that the parity of a permutation is unaffected by the choice of n in Definition 11.

● ● ●   **BREAK 8**

> Let σ be an arbitrary element of S_n and let ρ be the cycle $(\iota_1 \; \iota_2 \; \cdots \; \iota_r)$ in S_n. Show that $\sigma^{-1}\rho\sigma$ is the cycle $(\iota_1\sigma \; \iota_2\sigma \; \cdots \; \iota_r\sigma)$, where $m\sigma$ is the effect of the permutation σ on m.

Now it is plain that the set of even permutations of n objects is a *subgroup* of S_n; it is called the **alternating group on n objects** and written A_n. Moreover, if $n \geq 2$ and ρ is any *odd* permutation, then $A_n\rho$ is the set of *all* odd permutations. For certainly every permutation in $A_n\rho$ is odd; and, conversely, if σ is odd, then $\sigma\rho^{-1}$ is even, $\sigma\rho^{-1} \in A_n$, and $\sigma = (\sigma\rho^{-1})\rho$. Thus S_n is the disjoint union of the *two* cosets A_n and $A_n\rho$, consisting of the even and odd permutations, respectively, and hence A_n is a subgroup of S_n of index 2, as claimed in Section 2.

Notice that we could have argued above using ρA_n instead of $A_n\rho$; indeed, and remarkably, $\rho A_n = A_n\rho$. It is a special, and very important, property of some subgroups H of a group G that $gH = Hg$ for all $g \in G$. We call such subgroups **normal**. Our argument above can easily be generalized to show that every subgroup of index 2 of an arbitrary group G is normal.

● ● ●   **BREAK 9**

> (1) Let P be a subgroup of S_n and let Q be the subset of P consisting of the even permutations in P. Show that Q is a subgroup of P of index 1 or 2. Give an example of each possibility.
>
> (2) Show that, if H is a subgroup of G of index 2, then, for any $g \notin H$, we have $gH = Hg$.

Finally, let us remark here that the reason that even permutations figure so prominently in the discussion of symmetry in geometry is this: any symmetry of a polyhedron, and any orientation-preserving symmetry of a polygon, will induce an *even* permutation of the vertices. Try it and see!

● ● ●   **BREAK 10**

> (1) It is a fact that the regular tetrahedron (T) may be inscribed in the regular hexahedron or cube (H), as shown in Figure 9. Look at the rotations of the cube (from Break 6) that leave T occupying

		T	H	O	D	I
F		4	6	8	12	20
V		4	8	6	20	12
E		6	12	12	30	30
Axes	$\frac{2\pi}{2}$	?		6	?	
of	$\frac{2\pi}{3}$	?		4	?	
sym-	$\frac{2\pi}{4}$			3		
metry	$\frac{2\pi}{5}$				?	
Planes of				6	?	
symmetry				3		
Face angles		?		24	?	
Diagonals (interior)			4	3	60	?
					30	?
					10	

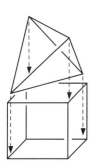

FIGURE 9

FIGURE 10 Here *T* is "tetrahedron," *H* is "hexahedron" (or "cube"), etc.

its original position within the cube. You should see that these rotations are simply the *even* permutations of the four strips of the diagonal cube (and that all the odd permutations move the tetrahedron from its original position to the same new position). Thus you have a dramatic confirmation that the symmetry group of *T* is A_4, the alternating subgroup of S_4, the symmetry group of *H*.

(2) Figures 10 and 11, which are about the Platonic solids (see Figure 4), were copied by JP from Pólya's personal notebook. Figure 11 is even in Pólya's own handwriting. See if you can fill in the blanks in Figure 10; and try to work out what Pólya means in Figure 11.

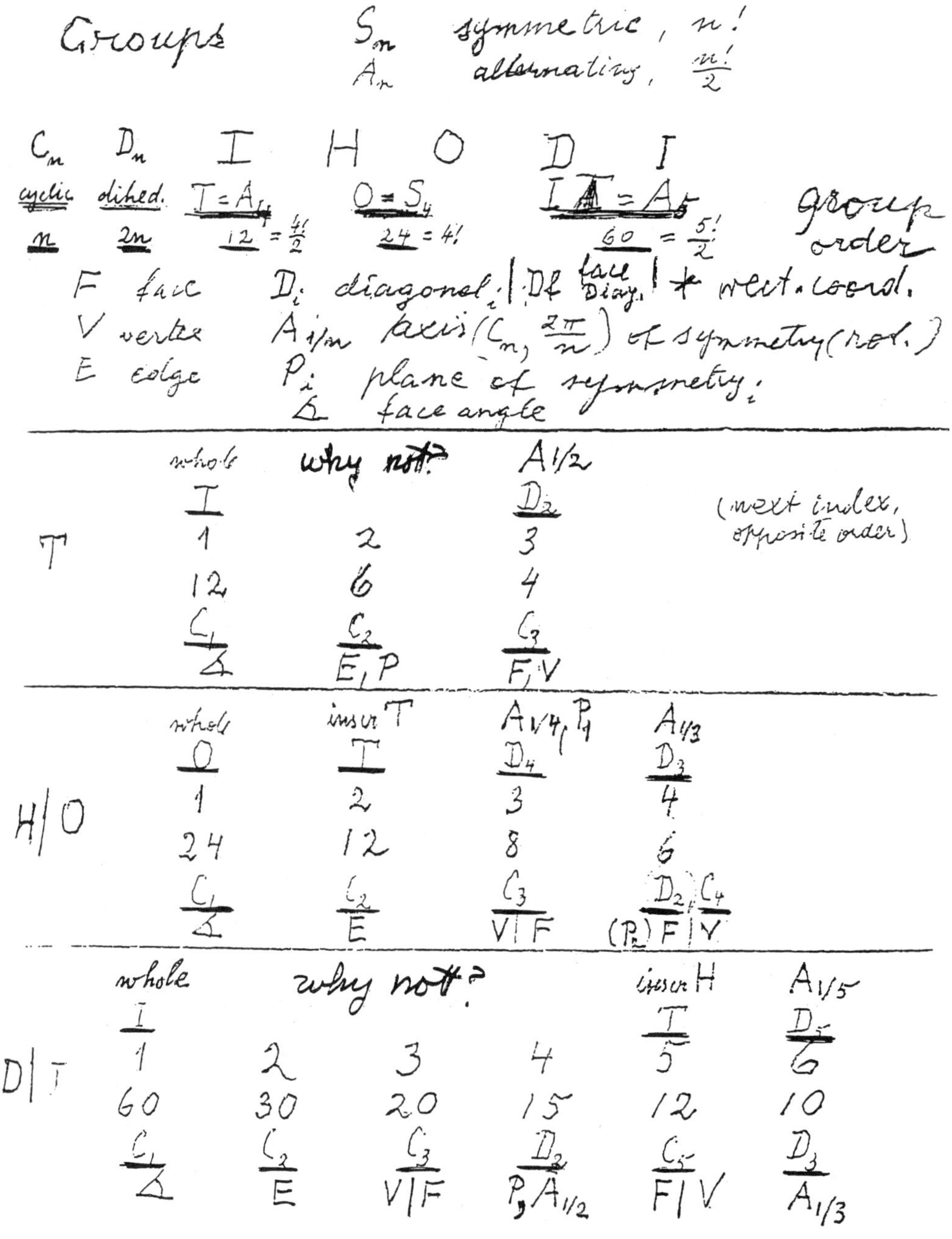

Groups S_n symmetric, $n!$
           A_n alternating, $\frac{n!}{2}$

C_n D_n I H O D I
cyclic dihed. $T = A_4$ $O = S_4$ $I...= A_5$ group
  n $2m$ $12 = \frac{4!}{2}$ $24 = 4!$ $60 = \frac{5!}{2}$ order

F face D_i diagonal; D_f face Diag. * vect. coord.
V vertex $A_{i/m}$ axis$(C_n, \frac{2\pi}{n})$ of symmetry(rot.)
E edge P_i plane of symmetry;
               $\angle$ face angle

T	whole	why not?	$A\,1/2$	
	I		D_2	(next index,
	1	2	3	opposite order)
	12	6	4	
	$\dfrac{C_1}{\angle}$	$\dfrac{C_2}{E,P}$	$\dfrac{C_3}{F,V}$	

H/O	whole	insur T	$A\,1/4, P_4$	$A\,1/3$
	O	T	D_4	D_3
	1	2	3	4
	24	12	8	6
	$\dfrac{C_1}{\angle}$	$\dfrac{C_2}{E}$	$\dfrac{C_3}{V\vert F}$	$\dfrac{D_2, C_4}{(P_2)F\vert V}$

$D\vert I$	whole	why not?			insur H	$A\,1/5$
	I				T	D_5
	1	2	3	4	5	6
	60	30	20	15	12	10
	$\dfrac{C_1}{\angle}$	$\dfrac{C_2}{E}$	$\dfrac{C_3}{V\vert F}$	$\dfrac{D_2}{P, A_{1/2}}$	$\dfrac{C_5}{F\vert V}$	$\dfrac{D_3}{A_{1/3}}$

FIGURE 11   As before, T is "tetrahedron," H is "hexahedron" (or "cube"), etc.

REFERENCES

1. Hilton, Peter, Derek Holton, and Jean Pedersen, *Mathematical Reflections — In a room with many mirrors*, 2nd printing, Springer Verlag NY (1997).

2. Hilton, Peter, and Jean Pedersen, *Build Your Own Polyhedra* Addison-Wesley, Menlo Park, CA, 1987, reprinted 1994, 1999.

3. Ledermann, Walter (chief editor), *Handbook of Applicable Mathematics, Supplement*, John Wylie & Sons, 1990.

4. Pedersen, Jean, and George Pólya, On problems with solutions attainable in more than one way, *College Math. J.* **15** (1984), 218–228.

5. Pólya, George, Intuitive outline of the solution of a basic combinatorial problem, *Switching theory in space technology* (ed. H. Aiken and W.F. Main), Stanford University Press, 1963, 3–7.

6. Pólya, George, *Mathematical Discovery — On Understanding, Learning, and Problem Solving* (Combined Edition), John Wiley and Sons, 1981.

7. Walser, Hans, *Symmetry*, (originally *Symmetrie*, B.G. Teubner, Stuttgart · Leipzig (1998)); English translation by Peter Hilton with the assistance of Jean Pedersen, Mathematical Association of America, 2001.

8. Willoughby, Stephen S., Carl Bereiter, Peter Hilton, and Joseph H. Rubinstein, *Mathematical Explorations and Applications*. SRA McGraw-Hill, 1998.

9 Parties

C H A P T E R

9.1 INTRODUCTION: CLIQUES AND ANTICLIQUES

We've probably all been to a party of some sort. And it always happens at a party that there are some people we know and some people we don't know. What's more, some of the people we know, know some of the people we know. Sometimes there's a group that goes around together. They all know each other. Such a group is often called a *clique*.

On the other hand, among the people we don't know there are people who don't know the people we don't know. In fact, there could be a group of people none of whom knows any of the others. This is a sort of *anticlique*. Of course, anticliques don't last for long. As soon as someone introduces himself or herself the anticlique gets broken up.

It may not have occurred to you, but there's a good chance that at any party you've been to there's been a clique or an anticlique of some size. What's the likelihood of this happening? Well, naturally, it all depends on the sizes of the cliques or anticliques. Perhaps once enough people get to the party you're bound to get any size clique or anticlique that you want.

Let's think about that for a minute. It might help if we call a clique that has r members who mutually know each other an *r-clique*. Similarly, an anticlique involving s people who all don't know each other we'll call an *s-anticlique*.

So how big does your party have to be in order for there to be an r-clique? You only need r people, provided they all know each other, so that's not much of a question. So how big does the party have to be to be sure that you absolutely *have* to have an r-clique? No, that's not a very good question either. If you keep inviting to the party people who don't know each other, you'll *never* get a clique at all. So let's try to rephrase the question. How big must the party be so that there is either an r-clique or an s-anticlique? That's not such a trivial question, is it?

• • • BREAK 1

How big does your party have to be to guarantee a 3-clique or a 3-anticlique?

Clearly, if you think about it, it makes sense to assume that r and s are at least 2. So what sort of a party has either a 2-clique or a 2-anticlique? Isn't 2 people enough to guarantee that? If you turn up at dinner with someone you know, there's a 2-clique. On the other hand, if it's a blind date, then there's a 2-anticlique.

Maybe this problem isn't going anywhere. What if we insist on either a 3-clique or a 2-anticlique? Will that produce a more interesting party? If we are going to have the ghost of a chance of getting a 3-clique there have to be at least three people at the party. So let's look at a party of three. If they all know each other, we have a 3-clique and we're finished. If there's not a 3-clique, then it's because at least two people don't know each other; and they then form a 2-anticlique. So then a three-person party forces either a 3-clique or a 2-anticlique.

• • • BREAK 2

How big is the party that forces either an r-clique or a 2-anticlique?

It may be a good idea to put this into graph-theoretical language. (Recall what we did in Chapter 5.) Suppose that we represent the people at the party by vertices. We'll join two vertices if the people they represent know each other. We won't join two vertices if the corresponding people don't know each other. An r-clique in this graph is a complete subgraph K_r, while an s-anticlique is a $\overline{K}_s$, the complement of K_s, that is, just s vertices. What we're trying to find is a number n such that every graph (party) on n vertices either has K_r as a subgraph or a set of S vertices with no edge joining them.

Actually we talked about coloring edges in Chapter 5. In terms of edge coloring what we're saying is this. Take K_n. Color its edges either red or blue, but don't worry if two red edges or two blue edges meet. How big does n have to be to ensure that K_n contains either a subgraph K_r with all red edges or a subgraph K_s with all blue edges? You can think of the red edges as joining pairs that know each other and the blue edges as joining pairs who don't know each other.

We know from Break 2 that if $r = 3$ and $s = 2$, then $n = 3$. A little bit of work will show that if $s = 2$ and r is any number (remember $r \geq 2$), then $n = r$.

• • • BREAK 3

What is the smallest value of n if $r = 2$ and $s \geq 2$?

Perhaps we can get somewhere by looking at $r = s = 3$. After all, if $r = 2$ or $s = 2$, then it's easy to find n. So now we're looking for the smallest value of n such that, however we color the edges of K_n red and blue, we always have to have either a red-edged K_3 or a blue-edged K_3. That is, however we color the edges, we have to have a monochromatic triangle.

The only way we have of tackling these problems at the moment is to try some numbers and see what works. It's no good starting at $n = 3$, though. If we don't have a red K_3, we don't necessarily have a blue K_3. For instance, we might have two red edges and one blue one.

What about $n = 4$? Figure 1 shows a coloring that doesn't produce a monochromatic triangle. Maybe that's not the only way either.

So let's try $n = 5$. Ah, that looks a little more promising. We've tried several edge colorings and they all give either a red or blue triangle. But how do we know we've covered all cases?

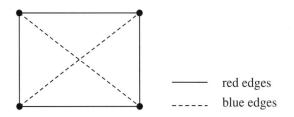

      ——— red edges

      ------ blue edges

FIGURE 1   No monochromatic triangle.

● ● ● **BREAK 4**

Does $n = 5$ for $r = s = 3$? Can you prove it? Can you find a counterexample?

Of course, the degree of every vertex in K_5 is four. So let's suppose that one of the vertices gets three red edges. (It looks as if, because of the symmetry between red and blue, we can start the argument with red edges without losing any generality. We'll come back to this in a moment.) So assume that v_1 is joined to v_2, v_3, and v_4 by red edges. If we have a red edge between any two of v_2, v_3, v_4, then we'll have a red triangle. If all of the edges between v_2, v_3, and v_4 are blue, then we'll have a blue triangle. So if any vertex has three red edges (or three blue edges), then we get a monochromatic triangle.

Unfortunately we are not done. It may be the case that every vertex has precisely two red edges. In this situation, we can break our K_5 up into two cycles C_5, one red and one blue (for the definition of a cycle, see Section 5.3). So there would be no red K_3 and no blue K_3, and so n has to be bigger than 5.

Well, try $n = 6$. Now the argument that we used when $n = 5$ may well be useful again. In K_6, every vertex has degree 5. This means that, at any vertex, there must be at least three red edges or at least three blue edges. (This is a very simple application of the pigeonhole principle.)[1] But now we are in business! Suppose v_1 is joined to v_2, v_3, v_4 with red edges. Then just one red edge between v_2, v_3, and v_4 will give our red K_3. On the other hand, if there are no red edges there, they must all be blue. In that case we are forced into a blue K_3. So the smallest n when $r = s = 3$ is given by $n = 6$.

● ● ● **BREAK 5**

Does $n = 6$ force only *one* monochromatic triangle?

In this last argument it's easy to see that we can choose *red* edges at v_1 without any real loss of generality. If we had chosen *blue* edges, we would simply have had to interchange red and blue throughout the argument.

---

[1] If you want to read more about the pigeonhole principle see [1, p. 251].

• • • **BREAK 6**

(1) So what is the smallest value of n if $r = 3$ and $s = 4$?

(2) How about when $r = s = 4$?

9.2 RAMSEY AND ERDŐS

The number n that we have been talking about is generally known as a Ramsey number. We define the ***Ramsey number***, $N(r, s)$, to be the smallest value of n such that if the edges of K_n are arbitrarily colored red and blue, then K_n contains either a red K_r or a blue K_s. So far, then, we've managed to determine a few of the smaller Ramsey numbers. That is, we know that

$$N(2, s) = N(s, 2) = s \quad \text{and} \quad N(3, 3) = 6.$$

It should also be clear by symmetry that

$$N(r, s) = N(s, r).$$

What perhaps is not so clear is that $N(r, s)$ exists at all. We'll move towards that as we go through this chapter.

But who was Ramsey, and why was he interested in his numbers? F.P. Ramsey was a logician who probably wasn't all that concerned about coloring the edges of graphs. It just so happened that some logical questions he was interested in could be interpreted combinatorially as we have done here. Unfortunately, he died at the young age of 26, of jaundice.

Ramsey was able to show in 1930 (see [8]) that his numbers actually do exist no matter what values r and s have. Indeed, we'll see later that the idea can be extended and the extended numbers will still exist.

In 1935, Ramsey numbers appeared again in quite a different setting in a paper by Paul Erdős and George Szekeres [3]. Szekeres was later to find one of the first few snarks (before Martin Gardner made them popular; see Section 5.9). He also went on to do a great deal of important work in his adopted Australia.

Let's digress now to talk about Erdős. Erdős, like Szekeres, was born in Hungary in 1913. (For more details on Erdős' life, see [6].) He went to university in Budapest, where he set out on a Ph.D. while still completing his undergraduate degree. From there he went on to make contributions to number theory and geometry, while virtually inventing combinatorial set theory by himself. His mathematical contribution of 1500 or more papers, results on various aspects of mathematics published in journals devoted to specialist mathematical areas, far exceeds that of any other mathematician alive or dead. Most of us are happy to publish 100 papers; 150 is quite an achievement, but 1500 is unparalleled.

Many of these publications had joint authorships with other mathematicians,[2] and this has precipitated the idea of a mathematician's Erdős number. You have an Erdős number of one if you have written a paper with Erdős. Anyone who hasn't written a paper with Erdős but who has written a paper with someone with an Erdős number of one, has an Erdős number of two; and so on. Another unique fact of Erdős's life was his wandering. He rarely stayed in any place for more than a month or so. In fact, his peripatetic nature may have contributed to mathematics more than his actual publications. As he moved from university to university, he carried unsolved problems with him and encouraged all whom he met to tackle them with him, sometimes offering monetary rewards for their solution. In this way he kept the mathematical cauldron boiling and influenced many young mathematicians to fly higher than they might otherwise have flown.

On his journeys around the world he often stayed with local hosts. Now, Erdős appeared to require little sleep, though it has to be said he took catnaps when the opportunity arose during the day. It was not unusual, though, for him to be up and thinking and working at two o'clock in the morning. This could be a difficulty for a host, who might suddenly awake in the middle of the night to find Erdős announcing that he had an idea that might solve the problem they had been working on a few hours ago.

Erdős' unworldly, naive, even childish manner could be a problem in the middle of the night, but it also fostered an extraordinary capacity for intellectual friendship. Many mathematicians had been helped by him during the course of his life, and there were those who were financially supported by him through difficult periods.

Another aspect of his naiveté was the language that he created. The connections are obvious when you think of "Sam" used for the United States and "Joe" for the USSR. But perhaps "slaves" for men and "bosses" for women are not as P.C. as they might have been, while "epsilons" for children may be a mystery to you if you haven't done any mathematical analysis yet. That he had strong opinions outside mathematics may be inferred from his use of the term "Supreme Fascist" for God.

Paul Erdős died in September 1996, a very well-loved and respected man whose family consisted of a significant portion of the world's mathematical research community. One of the important methods developed by Erdős

---

[2] Indeed, almost all — and almost all were actually drafted by his collaborators. This does not, however, by any means fully explain his prolific output.

was the ***probabilistic method***. This proves that certain things exist by showing that the probability of them not existing is zero. It's actually an unusual approach to proving things.

To give some idea of how the method works suppose that 12% of a sphere is painted red and 88% is painted blue. Is it possible to find 8 blue points on the sphere that are the vertices of a cube? Now, the probability of any vertex of a cube being red is 0.12. The probability that, of any two vertices of a cube, at least one is red, is approximately $0.12 + 0.12 = 0.24$. Continuing on like that, we see that the probability that, of all eight vertices of a cube, at least one is red, is $8 \times 0.12 = 0.96$. But 0.96 is less than one. So there must be a set of vertices of a cube that consists totally of blue points.

Naturally, the argument that $N(r, s)$ exists is more complex than this, but it follows the same line. The probability of the existence of various entities is established. Then we show that together the probability of the collection of all these entities is less than one. Hence the complement of the collection has to exist. We'll come back to this in Section 4.

9.3 FURTHER PROGRESS

So far we have discovered only that

$$N(2, s) = s \quad \text{and} \quad N(3, 3) = 6$$

Can we find other values of $N(r, s)$? What about $N(3, 4)$, for instance? How big does n have to be so that, when we arbitrarily color the edges of K_n red or blue, we either have a red triangle or a blue K_4.

Suppose $N(3, 4) = m$. Now choose an arbitrary vertex v in K_m and color r edges adjacent to v red, and the remaining b edges adjacent to v blue. We show this situation in Figure 2. Since deg $v = r + b$ and v belongs to K_m, we have $r + b = m - 1$. So $m = r + b + 1$.

Assume $r \geq 2, b \geq 3$. Now the r red edges coming out of v are adjacent to a set of vertices. Let these vertices generate a subgraph R. If we could guarantee a red edge in R or even a blue K_4, then we would have our red triangle or blue K_4. The blue K_4 is obvious. The red triangle is made up of two red edges coming from v and the one red edge in R. It's clear that, if $r \geq N(2, 4)$, we are sure to have either the red edge or the blue K_4 we need. Not to be sure is equivalent to $r \leq N(2, 4) - 1$.

Let's park that for a moment and let's look at the blue side of Figure 2. We define B in a similar way to that in which we defined R above. To

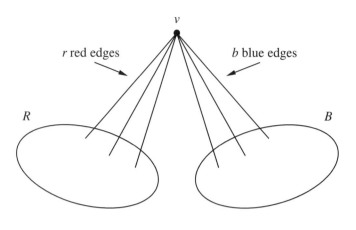

FIGURE 2

get a red triangle on the blue side would mean having a red triangle in B, whose vertices are adjacent to the blue edges coming from v. To get a blue K_4 in the graph, we'd only need a blue triangle in B. That triangle along with v and three edges coming out of v would give us the 4-clique we're after. To be sure that we have a red or blue triangle in B, it would be enough to require that $b \geq N(3, 3)$. Not to be sure is equivalent to $b \leq N(3, 3) - 1$. The question now is, how can we arrange things so that either $r \geq N(2, 4)$ or $b \geq N(3, 3)$? This fails if $r \leq N(2, 4) - 1$ and $b \leq N(3, 3) - 1$. But $m = r + b + 1$, so we must have $m > N(2, 4) - 1 + N(3, 3) - 1 + 1 = N(2, 4) + N(3, 3) - 1$ to use our argument. This shows that $N(3, 4) \leq N(2, 4) + N(3, 3) = 4 + 6 = 10$. But is $N(3, 4) = 10, 9, 8, 7, 6$?

• • • **BREAK 7**

Is it possible that $N(3, 4) < 9$?

What to do? When we were finding $N(3, 3)$, we worked upwards. We found a red/blue edge coloring of K_5, for instance, that had no monochromatic triangles. Can we find a red/blue coloring of K_7, K_8, or K_9 that has no red K_3 and no blue K_4?

To save a somewhat tedious search, look at the graph of Figure 3 (note that the only vertices are those around the outside marked with solid circles). This graph contains no triangles; hence there is no red K_3.

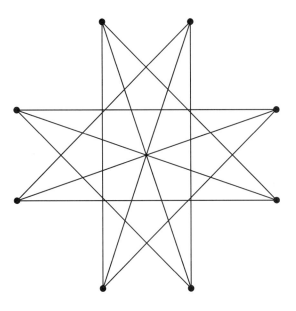

FIGURE 3   The red edges in K_8.

What happens if we color the remaining edges of K_8 in blue? Will we get a blue K_4? No. Any four vertices we choose already have at least one red edge between them. This means that $N(3, 4) > 8$. So

$$9 \leq N(3, 4) \leq 10.$$

But is $N(3, 4)$ equal to 9 or is it equal to 10?

Could $N(3, 4)$ possibly equal 9? Suppose not. Then we could color the edges of K_9 so that there is no red K_3 and no blue K_4. Suppose that we can color the edges of K_9 so that one vertex b, say, is adjacent to four edges colored in red and those four edges are adjacent to a set of vertices of size $4 = N(2, 4)$. Let this set generate the subgraph R. Now we're guaranteed to have a red edge in R or a blue K_4. We assumed that $N(3, 4) > 9$, so every vertex in K_9 cannot have more than 3 red edges.

On the other hand, if we have six blue edges adjacent to v, these six edges are adjacent to a set of vertices of size $6 = N(3, 3)$. Let this set generate the subgraph B. And now we're guaranteed to have a red triangle in B or a blue triangle in B. Either way we get our red K_3 or blue K_4 in K_9.

However, we assumed that $N(3, 4) > 9$, so every vertex in K_9 *has* to have at most 5 blue edges.

Now every vertex in K_9 has degree 8. To have at most 3 red edges, K_9 has to have at *least* 5 blue edges. But we know K_9 has to have at *most* 5 blue edges. Hence there are *exactly* 5 blue edges incident with each vertex of the graph K_9.

Actually that's something of a problem. This would mean that the graph on the blue edges would be regular of degree 5. We would have a graph on 9 vertices that is regular of degree 5. But no graph has an odd number of vertices of odd degree (see Chapter 5). Hence we have a contradiction. So $N(3, 4) = 9$.

• • • BREAK 8

So now try $N(3, 5)$ or perhaps $N(4, 4)$.

Armed with that success you might want to charge off and tackle $N(3, 5)$, $N(3, 6)$, or even $N(3, s)$. But before you do, think a bit. Can we generalize our earlier argument that $N(3, 4) \leq N(2, 4) + N(3, 3)$? With a little bit of care can we show that

$$N(r, s) \leq N(r - 1, s) + N(r, s - 1)? \tag{1}$$

Of course! The argument is exactly the same. You just have to work with r and s rather than with 3 and 4. Apart from that the idea is exactly the same.

Now not only can the relation (1) give us upper bounds for $N(r, s)$, given particular values of r and s, it can also prove the existence of $N(r, s)$. This existence is simply the result of double induction on r and s.

Right; so armed with the relation (1), it's clear that

$$N(3, 5) \leq N(2, 5) + N(3, 4) = 5 + 9 = 14$$

At this stage we have the same problem that we had when we discovered that $N(3, 4) \leq 10$. How much less than 14 is $N(3, 5)$? We adopt the same approach. Can we find a red/blue coloring of K_{13} or K_{12} or K_{11} or ... that yields no red 3-clique or blue 5-clique? If we can, then we can sandwich $N(3, 5)$ between this value and 14. Then with some subtle argument, we may be able to close up the gap.

Of course, all this is getting harder and harder. It turns out that $N(3, 5) = 14$. So, if you work hard, you *can* find a red/blue coloring of K_{13} such that there is no red K_3 or blue K_5.

Working on $N(3, 6)$ gives

$$N(3, 6) \leq N(2, 6) + N(3, 5) = 6 + 14 = 20$$

Now in this case it turns out that $N(3, 6) = 18$. To get this far you'll need to show that K_{17} won't play ball but K_{18} will. This all requires some effort, and the bigger the complete graph we have to work with, the greater the effort required. However, all Ramsey numbers that have been discovered so far have been found using the process we've talked about above. First find some upper bound for the number. Then stalk up on that number from below. This is generally done these days using a computer. Once your computer is having no luck producing the appropriately colored and sized cliques, then a more subtle argument has to be used to show that the Ramsey number is what you hoped it would be.

In this way the following $N(3, s)$ numbers have been determined:

$$N(3, 3) = 6 \qquad N(3, 4) = 9 \qquad N(3, 5) = 14$$
$$N(3, 6) = 18 \qquad N(3, 7) = 23 \qquad N(3, 8) = 28$$
$$N(3, 9) = 36$$

The only other Ramsey numbers that are known are

$$N(4, 4) = 18 \qquad N(4, 5) = 25$$

★ **9.4   N(r,r)**

In this section we give a probabilistic argument to show that $N(r, r) > n$, provided that $\binom{n}{r} 2^{1-\binom{r}{2}} < 1$. You may find this a little more difficult than the arguments of previous sections. The idea behind this method is to prove the existence of an object with a desired property by looking at the suitably defined random object. We then show that it has the required property with positive probability. Sometimes, the object itself is not actually presented.

To illustrate this method we present a result that Paul Erdős first proved in 1947. This result is the following theorem.

Theorem (Erdős, 1947):   *Suppose $\binom{n}{r} 2^{1-\binom{r}{2}} < 1$. Then $N(r, r) > n$.*

Proof   First of all, color the edges of K_n randomly in red and blue. This is done randomly by tossing a fair coin. As we move around K_n and meet an uncolored edge, we toss the coin. If the coin comes up heads, we color that edge red; otherwise, we color it blue.

Take any set S of r vertices. Let A_S be the event that the edges of S form a monochromatic K_r. Now the probability that one edge is red is $\frac{1}{2}$. The probability that two edges are red is $\frac{1}{2} \times \frac{1}{2} = \frac{1}{2^2}$. The probability that t edges are red is $\frac{1}{2^t}$. But K_r has $\binom{r}{2}$ edges. So the probability that all the edges of S are colored red is $1/2^{\binom{r}{2}} = 2^{-\binom{r}{2}}$.

Clearly we get the same probability if we're to have the edges of S form a blue K_r. Hence

$$\Pr(A_S) = 2^{-\binom{r}{2}} + 2^{-\binom{r}{2}} = 2^{1-\binom{r}{2}} \tag{2}$$

But S was just one set of r vertices. We need to sum $\Pr(A_S)$ over *all* possible sets of size r. What is $\Pr(A)$, where A is the set of all possible events like A_S? Finding an exact formula for $\Pr(A)$ is extremely difficult. However, we do know that

$$\Pr(A) \leq \sum \Pr(A_S),$$

where the sum is taken over all possible sets S of r vertices. But $\Pr(A_S)$ takes the same value $2^{1-\binom{r}{2}}$ for each of these sets. Moreover, there are plainly $\binom{n}{r}$ such sets S. Thus

$$\sum \Pr(A_S) = \sum 2^{1-\binom{r}{2}} = \binom{n}{r} 2^{1-\binom{r}{2}}$$

So

$$\Pr(A) \leq \binom{n}{r} 2^{1-\binom{r}{2}}$$

Now by the hypothesis of the theorem, $\binom{n}{r} 2^{1-\binom{r}{2}} < 1$. Hence $\Pr(A) < 1$.

Now consider $\overline{A}$, the complement of A. Since $\Pr(A) < 1$, $\Pr(\overline{A}) > 0$. And what is $\overline{A}$? Well, A is the event that there is *some* monochromatic K_r, so $\overline{A}$ is the event that there is no monochromatic K_r. Since $\Pr(\overline{A}) > 0$, there is an element in $\overline{A}$. This element is a coloring of K_n that has no monochromatic r-clique. We showed that $\Pr(\overline{A})$ was strictly positive by showing that, when we randomly colored K_n, $\Pr(A) < 1$. Since $\Pr(\overline{A}) > 0$, we got what we wanted, namely that $N(r, r) > n$. However, at no stage did we actually construct any red/blue colorings that have no red or blue r-cliques.

□

Just to see how good this bound is, let $r = 2$. We know that $N(2, 2) = 2$. But, by Erdős' Theorem, $N(2, 2) > n$ if $\binom{n}{2} 2^{1-\binom{2}{2}} < 1$. Now $\binom{n}{2} 2^{1-\binom{2}{2}} = \binom{n}{2}$. Since $\binom{2}{2} = 1$, we need $n < 2$ to satisfy the Erdős condition. But does $\binom{n}{2}$ exist for $n < 2$? (See [5] or Chapter 6 for a positive answer.) So $N(2, 2) > 1$; hence $N(2, 2) \geq 2$.

What about $N(3, 3)$? We know that $N(3, 3) = 6$. What value do we get for n that ensures that the expression $\binom{n}{3} 2^{1-\binom{3}{2}}$ is less than 1?

Now

$$\binom{n}{3} 2^{1-\binom{3}{2}} = \binom{n}{3} 2^{-2} = \frac{1}{4} \binom{n}{3} = \frac{1}{24} n(n-1)(n-2)$$

Can we solve the inequality $\frac{1}{24} n(n-1)(n-2) < 1$? Consider Table 1. Unfortunately, when $n = 4$ this expression equals 1. So Erdős' Theorem only tells us that $N(3, 3) > 3$.

$n =$	1	2	3	4	5	6
$\dfrac{1}{24} n(n-1)(n-2)$	0	0	0.25	1	2.5	5

Table 1

Do we do any better for $N(4, 4)$? When is $\binom{n}{4} 2^{1-\binom{4}{2}} < 1$? This is equivalent to asking, for what n is $\binom{n}{4} < 32$? The best we can do here is $n = 6$, so $N(4, 4) > 6$. Again this is a long way away from the 18 that we know is the best possible. The result does, however, give us a lower bound. The proof of the theorem also gives us a nice application of the probabilistic method.

• • • **BREAK 9**

Graph the function $y = \frac{1}{24} x(x-1)(x-2)$ for $-1 \leq x \leq 3$.

9.5 EVEN MORE RAMSEY

Mathematicians like to look for general patterns, so why stop with just 2 colors? What happens to K_n if we arbitrarily color its edges red, blue, and green? When can we guarantee a monochromatic triangle? When can we guarantee cliques of any size we happen to fancy. Just like $N(r, s)$, $N(r, s, t)$ exists. Coloring complete graphs in red, blue, and green, there is a smallest n such that K_n is forced to have a red K_r or a blue K_s or a green K_t. Let's see how this game is played for small values of r, s, and t.

Is it obvious that the game is easy for $r = s = t = 2$? In fact, it only starts to become interesting when $r = s = t = 3$, so we'll have a look at $N(3, 3, 3)$ and see what progress we can make.

From what we've done in Section 3, it might seem that $N(3, 3)$ is going to be a significant number in calculating $N(3, 3, 3)$. Look at Figure 4, where the vertex v is in some K_n and has $N(3, 3)$ edges colored red adjacent to it. These red edges meet a set of $N(3, 3)$ vertices, generating a subgraph R. (Does this ring any bells?) Now, if R contains a single red edge, then we have a red triangle, and we're finished. Otherwise, all of the edges between vertices in R are blue or green. Since R has $N(3, 3)$ vertices, we have to have a blue or a green triangle. So we're done!

But are we? If we are coloring the edges of K_n in red, blue, or green, how can we be sure that we have 6 red edges at v? Well, of course, we don't have to rely on red coming up 6 times. Any of the three colors would do. The argument would be the same regardless of the color.

In that case, how can we be sure that there are 6 edges at v of the same color? Perhaps if deg $v = 16$ we'll be home. We're applying the pigeonhole principle again. If we have 16 pigeons and three pigeonholes, aren't we forced to have one pigeonhole with 6 pigeons in it? And if deg $v = 16$, then v must be in K_{17}. So we have shown that $N(3, 3, 3) \le 17$.

So what about K_{16}? Surprisingly, K_{16} can be colored in three colors with no monochromatic triangle in sight. Hence $N(3, 3, 3) = 17$.

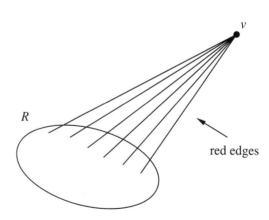

FIGURE 4 $|R| = N(3, 3)$

But now we're on a roll. Surely the argument above (which is similar to the one that produced $N(r, s) \leq N(r - 1, s) + N(r, s - 1)$) can now be extended to $N(3, 3, 3, 3)$ and $N(3, 3, 3, 3, 3)$ and so on. We need a new notation to be able to write this easily. Let $N(3, \cdots, 3)$ with k threes, be written $N(3_k)$. Then, in view of the case $k = 2$, we would expect that

$$N(3_{k+1}) \leq (k + 1)N(3_k) + 1$$

But why stick with threes? Surely, we can think of the Ramsey numbers $N(r_1, r_2, \cdots, r_k)$. The only problem is that they are impossibly difficult to calculate. Mathematicians are having enough trouble with $k = 2$. Only fairly recently was it shown that $N(4, 5) = 25$, and we only have bounds for $N(5, 5)$ and $N(4, 6)$. It turns out that $30 \leq N(3, 3, 4) \leq 31$, but we haven't got much further than that. Indeed, $N(3, 3, 3) = 17$ is the best result on Ramsey numbers with more than two colorings that we have.

To make more progress it appears that we need a new idea. Currently, we can find upper and lower bounds for the Ramsey numbers. Then the work is largely up to the computer to show that colorings don't exist from the lower bound up. And this requires an enormous amount of computer time.

There's plenty of room for new ideas here. Why don't you have a go?

9.6 BIRTHDAYS AND COINCIDENCES

You may not have noticed that 23 occurs all over the place. In basketball it was Michael Jordan's number as well as being the number of people on the soccer field when a game is in play. Twenty-three is also the number of problems that David Hilbert (1862–1943) proposed in his famous address in 1900. Hilbert was a very important mathematician of the period. Through his 23 problems he hoped to signpost the important areas of mathematical research in the twentieth century[3] (see [4]). And 23 is the number of letters in "Hilton, Holton, and Pedersen," but, much more importantly, it has to do with coincidental birthdays. What is the smallest number of people you have to have at a party in order that it's more likely than not that two of you have the same birthday? We will show that the answer to that question is 23. Of course, to show this, we'll have to prove that, with 23 people, the

---

[3]The Clay Mathematics Institute (CMI), of Cambridge, Mass., has issued a corresponding list of 7 problems for the new century, and has offered $1,000,000 for the solution of any one of them (see http://www.claymath.org).

probability of two having the same birthday is greater than $\frac{1}{2}$, while for a party with 22 people, the probability is less than $\frac{1}{2}$.

If you think about it very briefly, 23 seems awfully small. We suspect that most people would guess at a number more like 50 or 100. It should be clear that, if the party is big enough, then there must be two people present with the same birthday. We know that, forgetting about leap years, there are 365 days in a year. So a group of 366 *has* to have two people with the same birthday by the pigeonhole principle. How can we demonstrate that 23 is the magic number?

The first thing to notice is that the probability of event A, two people in a given group have the same birthday, and the probability of event B, no two people in the group have the same birthday, add up to 1. This is obvious, so

$$\Pr(A) + \Pr(B) = 1$$

It turns out that $\Pr(B)$ is easier to calculate than $\Pr(A)$. So we rearrange to get

$$\Pr(A) = 1 - \Pr(B).$$

The problem now[4] is to calculate $\Pr(B)$, the probability of event B. To do this we first suppose that there are 365 days in a year. Then person P_1 has her birthday on some given day, and, if B occurs, person P_2 has her birthday on one of the other 364 days. So the probability that P_1 and P_2 have different birthdays is $\frac{364}{365}$. For a two-person group, then, $\Pr(B) = \frac{364}{365}$.

What about a three-person group? Suppose P_3 doesn't have her birthday on the same day as either P_1 or P_2. There are 363 days available for P_3's birthday. So the probability of that happening is $\frac{363}{365}$.

This means that if the group size is three,

$$\Pr(B) = \frac{364}{365} \times \frac{363}{365}.$$

Continuing with this argument, the probability that no two people in a group of size n have a common birthday is given by

$$\Pr(B) = \frac{364}{365} \times \frac{363}{365} \times \frac{362}{365} \times \cdots \times \frac{365 - n + 1}{365}.$$

---

[4]To avoid the ugly his/her, let's suppose we're talking about groups of women.

So then we have

$$\Pr(A) = 1 - \frac{364}{365} \times \frac{363}{365} \times \frac{362}{365} \times \cdots \times \frac{366 - n}{365}.$$

If $n = 23$, $\Pr(A) \approx 1 - 0.4927 = 0.5173$. If $n = 22$,

$$\Pr(A) \approx 1 - 0.5243 = 0.4757$$

So for $n = 22$, $\Pr(A) < \frac{1}{2}$ and for $n = 23$, $\Pr(A) > \frac{1}{2}$. So parties of size 23 are the turning point. Once your party gets to this size it is more likely than not that there are two people with a common birthday.

● ● ●   **BREAK 10**

    (1) What is the probability that in a group of six, there are two people having their birthday on the same day of the week?

    (2) Find some more occurrences of the number 23. Are there more of those than you would expect?

9.7   COME TO THE DANCE

We began to look at Ramsey numbers because we wanted to know how big a party we'd need to have to be sure that there was either an r-clique or an s-anticlique. Well now, everyone's arrived at the party, the carpet has been rolled up, the music's on, and it's dancing time. But wait! How can we be sure that everyone gets to dance? That's no problem. If there are an equal number of girls and boys, they'll all just pair up and away we go.

But what if girls decide that they'll only dance with a boy they already know? Can we still get everyone dancing?

Is that condition likely to cause problems? It doesn't seem much of a restriction. Oh, but what if 15 boys and 15 girls came to the party and what if the 15 girls altogether only knew 11 of the boys. Then not all the girls would get to dance — unless they decided that they would dance with a boy they didn't know after all.

But is that the only problem they could run into? In the same way, let's assume that you find 8 girls who together only know 6 boys. The other 7 girls might know 9 boys. They would be OK. But you wouldn't be able to pair up the 8 girls with guys they knew.

So in order for there to be a chance to get the dancing going, any subset of the girls needs to know at least the same number of boys. But does it work the other way round? If every subset of girls knows at least the same number of boys, will we be able to pair all of the girls off with boys they know?

Apart from being a problem that can be solved by looking at graphs, this problem has another connection with Ramsey numbers. The result we are about to produce was proved in 1936 when Erdős and Szekeres were having their success. But while Erdős and Szekeres were working in Hungary, the person who was worrying about the dancing problem was in England, at Cambridge University to be precise. And his name was Philip Hall.

The first question that we want to tackle is how to translate boys and girls into graphs and what kind of a graph do they make? This isn't too difficult. Here we're worried about boys and girls dancing and girls knowing boys. So we'll let the girls form a subset G of the vertices, and we'll let the boys correspond to the remaining vertices B. In this graph with vertex set $G \cup B$, we'll only join a vertex representing a particular girl, g, to a vertex representing a particular boy, b, if the girl g knows the boy b.

Let's have a look at the graph of Figure 5. This graph represents the following situation:

Girls	Boys they know
g_1	b_1, b_2, b_5
g_2	b_1, b_2
g_3	b_3, b_4, b_5
g_4	b_2, b_3, b_5, b_6
g_5	b_1, b_4, b_6

Right from the start it's clear that one boy is about to sit this one out. But can we get a dance for all the girls, always assuming that they dance

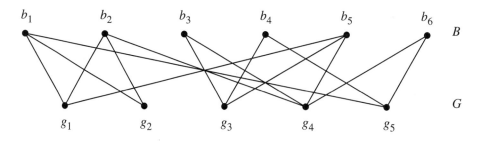

FIGURE 5

with boys they know? Let's try to do some pairing. So we see that we can pair g_1 and b_1, g_2 and b_2, g_3 and b_4, g_4 and b_5, and g_5 and b_6.

As far as the graph is concerned what's going on? Well, first of all, what sort of graph have we got? Graphs in which the vertex set can be divided into two sets X and Y so that the only edges in the graph are between vertices in X and vertices in Y are called ***bipartite*** graphs (see Section 5.7). The sets X and Y are called the ***parts*** of the vertex set. The graph in Figure 5 is bipartite, and the parts are B and G. All girl-boy graphs of the type we are considering are bipartite. We only have edges between G and B.

So how can we describe the pairing? In the bipartite graphs, we are interested in sets of distinct edges which have no end vertices in common. They are called sets of ***independent*** edges. We don't want a boy or a girl to be dancing with two partners simultaneously.

Such a set of independent edges is usually called a ***matching***. But we really want a special type of matching. We want a matching such that every girl belongs to one end of an edge in the matching. We say that such a matching ***saturates*** the set G.

Let's be more formal with our statement that "every subset of girls knows at least the same number of boys." Let H be a graph and let u be a vertex of H. Write $N(u)$ for the set of vertices of H adjacent to u. We call $N(u)$ the ***neighborhood of the vertex*** u. Extending our notation to subsets S of the set of vertices of H, let[5] $N(S) = \bigcup_{u \in S} N(u)$. Then $N(S)$ is called the ***neighborhood of the set*** S.

Now let $|N(S)|$ denote the size of the neighborhood of S. For example, in Figure 5, let $S = \{g_1, g_2, g_4\}$. Since $N(g_1) = \{b_1, b_2, b_5\}$, $N(g_2) = \{b_1, b_2\}$, and $N(g_4) = \{b_2, b_3, b_5, b_6\}$, then

$$N(S) = \{b_1, b_2, b_3, b_5, b_6\}$$

and $|N(S)| = 5$.

If S is a set of girls, then surely the set of boys they know is precisely $N(S)$. So if the girls know at least as many boys, $|N(S)| \geq |S|$. What we are beginning to conjecture is the following:

Conjecture   *There is a matching that saturates the set of girls in a bipartite graph if and only if $|N(S)| \geq |S|$ for every subset S of the set G of girls.*

---

[5]The symbol $\bigcup$ means the ***union*** (of the sets $N(u)$).

• • • BREAK 11

(1) Check the conjecture for the following party:

Girls	Boys they know
g_1	b_1, b_2, b_3
g_2	b_1, b_3, b_4, b_5
g_3	b_4, b_5

(2) Check the conjecture for the following party:

Girls	Boys they know
g_1	b_1, b_2, b_4, b_5
g_2	b_1, b_2, b_3
g_3	b_4, b_5
g_4	b_4, b_5
g_5	b_4, b_5

9.8 PHILIP HALL

There is a lovely proof of the correctness of the conjecture, due to the eminent Cambridge group theorist Philip Hall. We are now going to give you that proof in a particular but not special case.

Often one part of the proof of an "if and only if" statement is quite straightforward. Such is the case here. Clearly, if there is a matching that saturates the set of girls, then any subset of the girls, no matter of what size, must know that many boys. In fact, it's possible that those girls know more boys. So, $|S| \leq |N(S)|$ for every subset S of girls.

So it's the other way round that's a problem. How can we show that, if $|N(S)| \geq |S|$ for every subset S of G, then we can arrange the appropriate dancing partners? Of course, when you think about it, this is a very unadventurous group of girls who are not going to meet any nice new fellows if they stick to their guns and won't dance with any boys they don't know. But that's irrelevant right now.

If we can't see how to settle this half of the conjecture directly, maybe we can sneak up on it. Well, perhaps we can do it if there is only one girl? Oh yes, that's not too bad. If $S = \{g\}$, then $|S| = 1$ and $|N(S)| \geq 1$. So there is at least one boy that young g knows, and they can go off and dance.

With that little success we can try $|S| = 2$. Let $S = \{g_1, g_2\}$, say. Now $|N(S)| \geq 2$. Surely, then, we can pair up g_1 and g_2? But what if $|N(S)| \geq 2$ because g_1 knows two boys? Maybe g_2 doesn't know any? No, that can't be. After all, $|\{g_2\}| \leq |N(\{g_2\})|$, so g_2 has to know at least one chap.

It's beginning to look as if we might be able to get through this. However, by the time we get to 15 girls, there might be a large number of subcases to check, subcases like that of g_2 above, and worse!

Let's take a big jump. Let's try to do it for a set G of 15 girls, using the assumptions of the conjecture, and let us assume that we can do it for all sets of 14 girls or fewer.

Now it's just possible that not only is $|N(S)| \geq |S|$ but that $|N(S)| \geq |S| + 1$ for *every* subset S of G here. The reason we've added that 1 above is to make it easy to do this next step. That step is to pair up some girl, any girl, with a boy b, that she's happy to dance with. As the result of this we may reduce $N(S)$ for any S we choose now, but we will only reduce $N(S)$ by 1 — the boy b. So among the 14 girls still looking for a dancing partner, we'll still have $|N(S)| \geq |S|$. A little while ago we assumed that this would give us dancing partners for 14 of the girls. So then we pair up the remaining girls and give a sigh of relief.

That is, until we realize that up above we added a very strong condition, $|N(S)| \geq |S| + 1$, for all subsets S of G. Unfortunately, it might be the case that, for some number $n < 15$, there is a subset S' of G such that $|N(S')| = |S'| = n$. What to do then? Suppose, for example, we have S' such that $|N(S')| = |S'| = 6$. Well, the first thing to note is that the critical $|N(T)| \geq |T|$ condition will hold for all subsets T of our S'. So we can pair up the members of S' with 6 boys of their choice. Now remove S' and their chosen partners from the graph and look at what is left. So we still have 9 girls not paired. To make things quite clear let us use the notation $\tilde{N}(S)$ for the neighborhood of S in the *deleted* graph. Naturally, if the $|\tilde{N}(S)| \geq |S|$ condition holds for all S in the graph we have left, then they can go off and dance. But does that condition hold?

Suppose it doesn't. Suppose somewhere, after pairing up the 6 girls of S' and removing them from the graph, there is a subset U of the 9 remaining girls, where $|\tilde{N}(U)| < |U|$. What then? Here's an idea. What does that tell us about S' and U? If we put them together,

$$|N(S' \cup U)| \leq |N(S')| + |\tilde{N}(U)| < |S'| + |U| = |S' \cup U|$$

since S' and U have no girls in common. So here we have a set $V = S' \cup U$ for which $|N(V)| < |V|$. But that can't be! It's against our original assumption. Hence every subset U of the 9 girls has the property $|\tilde{N}(U)| \geq |U|$.

Since $9 < 14$, we know that we can pair up the 9 girls with the remaining boys.

We paired up the 9 girls, we'd already paired up the 6 girls, so all 15 have a partner! When you look at all that more closely you might see that 15 was really a red herring. It could just have easily been 150, 1500, or 15 million. The principle would have been the same. Doesn't that sound like a proof by induction?

Notice that what we have done above is to adopt an expository procedure that we recommend in [5] (see p. 327). We have applied the principle of considering a Particular but not Special Case.

• • • BREAK 12

Prove the difficult part of the dancing conjecture.

And that gives you Hall's Theorem.

Hall's Theorem   *Suppose the girls at a party know some boys and will only dance with boys they know. Then the girls can be paired up to dance with boys of their choice if and only if, for any subset of the girls, the number of boys they know is at least as big as the number of girls in the subset.*

9.9   BACK TO GRAPHS

In terms of graphs, Hall's Theorem can be restated as follows.

Hall's Theorem   *Let Q be a bipartite graph with bipartition $VQ = X \cup Y$, where VQ is the set of vertices of Q. Then Q contains a matching that saturates X if and only if $|N(S)| \geq |S|$ for every subset S of X.*

If we go back to Figure 5 and look at all possible 32 subsets of $\{g_1, g_2, g_3, g_4, g_5\}$, we can check the $|N(S)| \geq |S|$ condition. It might take a bit of time, but we can do it. On the other hand, we can see that g_1b_1, g_2b_2, g_3b_3, g_4b_5, and g_5b_4 gives us a saturated mapping.

This rather suggests that the $|N(S)| \geq |S|$ condition is pretty hard to apply in practice. As it turns out, however, it does have its uses. Let's look at one of these in particular. Suppose Q is an r-regular bipartite graph. That is, suppose every girl is prepared to dance with any one of exactly r boys and every boy is on r of the girls' dance lists. The graph $K_{3,3}$ is a 3-regular

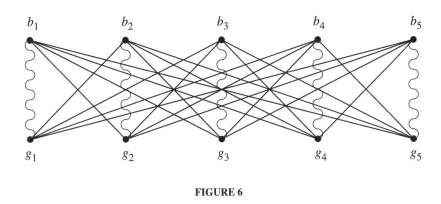

FIGURE 6

complete bipartite graph, $K_{m,m}$ is an m-regular complete bipartite graph, and C_{2n} is a 2-regular bipartite graph.

Fine, so Q is r-regular and bipartite. It turns out that not only can every girl get her dance with a boy of her choice, but every boy also gets to dance. That's some sort of bonus, at least for the host and hostess, who don't want the odd dance-less boy attacking the supper before eating time. This kind of matching, where every vertex gets saturated, is called a ***perfect matching***. In Figure 6 we show a perfect matching in $K_{5,5}$. Indeed, g_1b_1, g_2b_2, g_3b_3, g_4b_4, g_5b_5 is one of many perfect matchings in $K_{5,5}$.

The first thing to notice about r-regular graphs is that in the bipartition $X \cup Y$ of the vertices, $|X| = |Y|$.

● ● ●   **BREAK 13**

Can you show that $|X| = |Y|$ in an r-regular bipartite graph with parts X and Y?

So can we show that Hall's crucial condition, namely that $|N(S)| \geq |S|$ for every subset S of X, holds in an r-regular bipartite graph B? Take some subset $S \subseteq X$. Let E be the set of edges coming out of S. (See Figure 7.) Because every vertex of Q has degree r, $|E| = r|S|$. On the other hand, let E' be the set of edges incident to $N(S)$. We now have two things to report. The first of these is that $|E'| = r|N(S)|$. That follows in the same way that we got $|E| = r|S|$. And the second thing is that $E \subseteq E'$, as we can see from Figure 7. So $|E| \leq |E'|$.

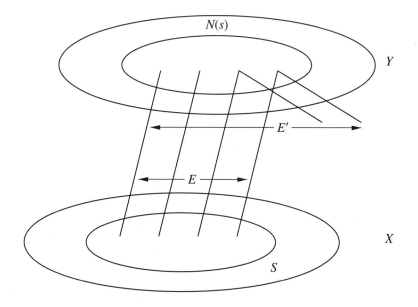

FIGURE 7

So putting all that together we get

$$|N(S)| = \frac{1}{r}|E'| \geq \frac{1}{r}|E| = |S|$$

This holds for every subset of X. Hence, by Hall's Theorem, we have a perfect matching in B.

● ● ● **BREAK 14**

Show that this goes even further. An r-regular bipartite graph not only has a perfect matching, but it has r perfect matchings. What's more, these r perfect matchings are all disjoint, but together they include all the edges of the graph.

At this point you might be wondering, why bother? And you might be wondering this for two reasons. First, if the girls can't sort out their own dancing partners, then perhaps they don't deserve to be saved by a 1935 theorem. Second, why should we worry about matching things up anyway?

There's not much point in fussing over the first reason, but, as to our second question, it turns out that matching is important in some industrial applications. For instance, suppose you have a work force and a collection of jobs to be done. Maybe not all of the workforce can do all of the jobs. So you might want to check out the matching business either to make sure that all your workforce is employed at a job or even that every job is being done by someone. This is a matching problem of the type that Philip Hall was good at.

Actually you can take this even further if you know how long each person takes to do each job. In that case you not only want a matching but a most efficient matching. But we'll leave you to chase that up for yourself.

9.10 EPILOGUE

In this chapter we've climbed some of the high points of graph theory. We stopped as the mists started to swirl in. As far as Ramsey Theory goes, it's clear that we've still got a long way to go. All of the last Ramsey numbers to be found required a considerable amount of computer time to produce. We could get more numbers if we had faster computers. For mathematicians, as we said in Chapter 5, that's a sorry state of affairs. We'd very much like to find some "nice" approach. It would be good if we could find some better approximations so that we would not need so much number-crunching.

In the meantime, in the traditional way of mathematicians when they hit a virtual brick wall, extensions of the problem have been made. Instead of asking for monochromatic cliques they have tried more modest goals. If we color the edges of K_n arbitrarily in red and blue, how big must n be in order to get a monochromatic cycle? How big must n be to have a red triangle and a blue $K_{1,3}$? About now you should be able to see that you can invent your own problems. You might even be able to solve some of them.

So Ramsey Theory is looking for some new ideas. Maybe they'll come through looking at things other than monochromatic cliques. In the meantime, the calculation of Ramsey numbers is in a similar state to that of the attempted proof of the Four Color Conjecture in the 1960s. Gradually, maps with an increasing number of countries were being shown to be 4-colorable. Now, year by year, someone manages to find yet another Ramsey number. But progress is slow and hard won.

Our other problem of the chapter, looking for matchings, preferably perfect matchings, in graphs is also receiving attention. There are some big results in the field such as Hall's Theorem, a theorem of Tutte, more recently the Edmonds–Gallai Theorem (see [7]), and so on. Where are people heading in this area of mathematics?

A number of things are on the go. For instance, for a long time people have been thinking about matchings with cycles. These are called *2-factors*. They are subgraphs of degree 2 (hence cycles) that include all the vertices of the graph under consideration. (When you think about it, perfect matchings are subgraphs of degree 1 that include all the vertices of the graph.) Once you've started to think about matchings that way, you soon come up with k-factors — subgraphs of degree k that cover the graph. As long ago as 1898, Petersen (his graph has a central place in Section 5.10) came up with a fundamental result in this area.

Theorem (Petersen) *Let G be a connected graph that is regular of degree 3. Suppose G has at most two bridges. Then G has a 2-factor.*

Here a *bridge* is an edge whose removal disconnects the graph. Note that, because the graphs of the theorem are regular of degree 3, the existence of a 2-factor is equivalent to the existence of a perfect matching. Hence this theorem also explains why the graph of Figure 8 doesn't have a perfect matching: it has three bridges, 12, 13, 14.

There are many other areas of matching theory, and areas related to matching theory, that we haven't been able to mention here. An interesting and valuable introduction to the subject can be found in the preface of [7]. If you become interested in this topic you should, of course, read further than just the preface.

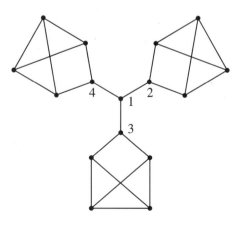

FIGURE 8

• • • **BREAK 15**

(1) Find $N(2, 2, 2)$, $N(2, 2, 3)$, and $N(2, 3, 3)$.

(2) Prove that $N(m, n) \leq \binom{m+n-2}{m-1}$, $m, n \geq 2$. (Hint: See item (1) of Section 3.)

(3) Let G be a graph that is regular of degree 4 and has an even number of vertices. Let $G - \{v\}$ be the graph obtained from G by deleting the vertex v and all edges incident with it. Show that, for any v, $G - \{v\}$ contains at most one component with an odd number of vertices. Must such a graph have a perfect matching?

REFERENCES

1. Erdős, Paul, and George Szekeres, A combinatorial problem in geometry, *Compositio Math.* **2** (1935), 463–470.

2. Erdős, Paul, Some remarks on the theory of graphs, *Bull. Amer. Math. Soc.* **53** (1947), 292–294.

3. Hall, Philip, On representations of subsets, *J. London Math. Soc.* **10** (1935), 26–30.

4. Hilbert, D., Mathematical Problems: Lecture Delivered Before the International Congress of Mathematicians at Paris in 1900, trans. Mary F. Winston, *Bulletin of the American Mathematical Society* **8** (1902), 437–479.

5. Hilton, Peter, Derek Holton, and Jean Pedersen, *Mathematical Reflections — In a Room With Many Mirrors*, 2nd printing, Springer Verlag NY, 1998.

6. Hoffman, Paul, *The Man Who Loved Only Numbers*, Fourth Estate, London, 1998.

7. Lovász, László, and Michael D. Plummer, *Matching Theory*, Akadémia: Kiadó, Budapest, 1986.

8. Ramsey, F., On a problem of formal logic, *Proc. London Math. Soc.* **30** (1930), 264–286.

Information regarding Ramsey Theory can be found on the web at http://www.combinatorics.org/surveys/index.html

Selected Answers to Breaks

NOTE: Where the symbol $\boxed{V}$ appears before an answer it indicates that answers may vary.

CHAPTER 1

1. The answers are $\frac{1}{2}$; $\frac{1}{3}$; yes. We are dealing here with *infinite* sets, so the probability of an integer belonging to some subset depends on the *density* of that subset and not merely on its cardinality. After all, what is the probability that an integer chosen at random is divisible by 1? But there are as many even integers as integers.

3. $\boxed{V}$ (1)

A	5	5	5	5	1	1
B	4	4	4	4	4	4
C	4	4	4	4	3	3
D	6	6	2	2	2	2

$\boxed{V}$ (2) *A* is best in our example. Just calculate the expectations of *A* and *B*, the only two serious contenders.

$\boxed{\text{V}}$(3)

A	5	5	5	5	1	1
B	5	5	5	4	1	1
C	4	4	4	4	4	4
D	4	4	4	4	3	3
E	6	6	2	2	2	2

5.   (1) The "effective monkey count" is now $\sum_{j=1}^{m} n_j$ instead of m. The problem is the same.

    (2) Here d plays the role of $\frac{s-1}{s}$, $s = \frac{1}{1-d}$, $s - 1 = \frac{d}{1-d}$. Thus $v_0 = \frac{v_i}{d^i}$, $v_i = d^i v_0$, $u_i + \frac{dm}{1-d} = d^i (u_0 + \frac{dm}{1-d})$,

$$u_i = d^i u_0 + \frac{dm}{1-d}(d^i - 1) = d^i u_0 - \frac{d^i - 1}{d-1}\, dm.$$

CHAPTER 2

1. Let f be the number of 4¢ stamps and t the number of 10¢ stamps. Then

$$4f + 20f + 10t = 200.$$

So

$$24f + 10t = 200,$$
$$24f = 10(20 - t),$$
$$12f = 5(20 - t).$$

    Since 5 is a prime that divides the RHS, it must divide the LHS. Since 5 does not divide 12, it must divide f. Hence $f = 5\bar{f}$. Then

$$12\bar{f} = 20 - t.$$

The only possible value of $\bar{f}$ is 1. So $\bar{f} = 1$, $f = 5$, and $t = 8$. This means that Dennis bought five 4¢ stamps, fifty 2¢ stamps, and eight 10¢ stamps.

3. The most likely solution is $x = 5$, $y = 3$. (See Section 8.)

5. $$(u^2 - v^2)^2 + (2uv)^2 = u^4 - 2u^2v^2 + v^4 + 4u^2v^2$$
$$= u^4 + 2u^2v^2 + v^4$$
$$= (u^2 + v^2)^2$$

So $(u^2 - v^2, 2uv, u^2 + v^2)$ is indeed a Pythagorean triple.

7. For any number n, $n \equiv 0, 1, 2,$ or 3 (mod 4), so $n^2 \equiv 0, 1, 4,$ or 9 (mod 4). But this means that $n \equiv 0, 1, 0,$ or 1 (mod 4).

9. (1) Let $p, g,$ and s be the number of pigs, goats, and sheep, respectively. Then

$$p + g + s = 100$$
$$21p + 8g + 3s = 600$$

Eliminating s gives $18p + 5g = 300$. Since 5 and 300 have a factor of 5 and 18 doesn't, 5 is a factor of p. Let $p = 5a$. Then

$$18a + g = 60$$

Similarly, 6 divides g. Let $g = 6b$. Then

$$3a + b = 10$$

But p has to be even, so a is even. If $a = 4$, b is negative. So $a = 2$, $b = 4$. Hence $p = 10$, $g = 24$, and $s = 66$.

(2) Work modulo 5 here (see table below).

m (mod 5)	0	1	2	3	4
m^2 (mod 5)	0	1	4	4	1
m^4 (mod 5)	0	1	1	1	1

Hence the only fourth powers modulo 5 are 0, 1.

Now, $251 \equiv 1$ (mod 5). Hence $x^4 + 251 \equiv 1$ or 2 (mod 5). On the other hand, $5y^4 \equiv 0$ (mod 5). Hence $x^4 + 251$ can never equal $5y^4$ if x and y are required to be integers.

[For what values of k does $x^4 + 251 = ky^4$ have solutions?]

(3) Approach this in a similar way to the Pythagorean triple argument.

Case 1. Suppose x and y are both even. Then x^2 and y^2 are both divisible by 4. Clearly, $80z + 102$ is not divisible by 4.

Case 2. Suppose x is even and y is odd. (A similar argument applies for x odd and y even.) Then y^2 is odd. So $x^2 + y^2$ is also odd. Clearly, $80z + 102$ is even.

Case 3. Suppose x and y are both odd. Now, the square of an odd number is congruent to 1 (mod 8). This is because $(2n + 1)^2 = 4n(n + 1) + 1$, and $n(n + 1)$ is obviously even. Thus $x^2 + y^2 \equiv 2$ (mod 8). But $80z + 102 \equiv 6$ (mod 8). Hence $x^2 + y^2 = 80z + 102$ has no integral solutions.

(4) Suppose that Diophantus lived to be d years old. Then he

was a boy for $\dfrac{d}{6}$ years;

had to shave after $\dfrac{d}{12}$ more years;

was married after a further $\dfrac{d}{7}$ years;

had a son 5 years later;

his son died $\dfrac{d}{2}$ years later; and

then he died 4 years later.

So $\quad d = \dfrac{d}{6} + \dfrac{d}{12} + \dfrac{d}{7} + 5 + \dfrac{d}{2} + 4$

$\qquad = \dfrac{75}{84} d + 9$

Thus $\dfrac{9d}{84} = 9$, so $\dfrac{d}{84} = 1$.

Hence $d = 84$. Diophantus lived to the very good age of 84 years.

(5) If $x = \dfrac{a}{N}, y = \dfrac{b}{N}, z = \dfrac{c}{N}$ is a solution (we can always represent the three rational numbers as fractions with a common denominator), then so is $x = a, y = b, z = c$.

(6) Let $n \neq 1, 2, 4$. If n is even, then we know by FLT that there is no solution in positive integers. So suppose n is an odd integer, $n \geq 3$, and suppose $x^{n/2} + y^{n/2} = z^{n/2}$ has a solution. Let (x, y, z) be a solution with minimum z. Now, $x^n + 2x^{n/2} y^{n/2} + y^n = z^n$, so $2x^{n/2} y^{n/2}$ is an integer. Since its square is even, it must be even, so $x^{n/2} y^{n/2}$ is an integer. We prove that x, y are coprime. For if not, let p be a prime such that $p|x$, $p|y$. Then $p^2|x^n y^n$; but $x^n y^n$ is a square, so $p|x^{n/2} y^{n/2}$. But then $p|z^n$, so $p|z$. We would have $x = px_1$, $y = py_1$, $z = pz_1$, and (x_1, y_1, z_1) would be a solution with $z_1 < z$, contradicting the minimality. Hence x, y are coprime, so x^n, y^n are coprime. Thus, since $x^n y^n$ is a square, each of x^n, y^n is a square. But since n is *odd*, this means that x, y are squares, say $x = a^2, y = b^2$. It follows that $z^{n/2} = a^n + b^n$, so $z^{n/2}$ is an integer, so z^n is a square. Again it follows that z is itself a square, say $z = c^2$, and $a^n + b^n = c^n$, contradicting FLT.

CHAPTER 3

1. (1) $L_{n+1}L_n = (\alpha^{n+1} + \beta^{n+1})(\alpha^n + \beta^n)$

$\qquad = \alpha^{2n+1} + \alpha^{n+1}\beta^n + \alpha^n\beta^{n+1} + \beta^{2n+1}$

$\qquad = (\alpha^{2n+1} + \beta^{2n+1}) + \alpha^n\beta^n(\alpha + \beta)$

$\qquad = L_{2n+1} + (-1)^n, \qquad \text{since} \quad \alpha\beta = -1 \quad \text{and} \quad \alpha + \beta = 1.$

$$F_{n+1}L_n = \frac{\alpha^{n+1} - \beta^{n+1}}{\alpha - \beta}(\alpha^n + \beta^n)$$

$$= \frac{\alpha^{2n+1} - \alpha^n\beta^{n+1} + \alpha^{n+1}\beta^n - \beta^{2n+1}}{\alpha - \beta}$$

$$= \frac{\alpha^{2n+1} - \beta^{2n+1}}{\alpha - \beta} + \frac{\alpha^n\beta^n(\alpha - \beta)}{\alpha - \beta}$$

$$= F_{2n+1} + (-1)^n, \qquad \text{since} \quad \alpha\beta = -1.$$

(2) Since the Fibonacci and Lucas sequences *start* the same mod 2, they *are* the same mod 2. Thus we need only look at the one sequence mod 2; it starts

$$01101\cdots$$

Since 01 has repeated, it repeats 011 (with period 3).
    Thus

$$F_n \equiv 0 \bmod 2 \iff L_n \equiv 0 \bmod 2 \iff 3|n$$

3. (1) (i) $L_{18} = 5778 \equiv 2 \bmod 16$, or, calculating mod 16, we have

L_{12}	L_{13}	L_{14}	L_{15}	L_{16}	L_{17}	L_{18}
2	9	11	4	15	3	2

(ii) Calculating mod 49,

L_{12}	L_{13}	L_{14}	L_{15}	L_{16}	L_{17}	L_{18}	L_{19}	L_{20}	L_{21}	L_{22}	L_{23}	L_{24}
28	31	10	41	2	43	45	39	35	25	11	36	47

so

$$L_{24} \equiv -2 \bmod L_4^2$$

(2)
$$L_{2(k-1)r} L_{2r} = \left(\alpha^{2(k-1)r} + \beta^{2(k-1)r}\right)\left(\alpha^{2r} + \beta^{2r}\right)$$
$$= \alpha^{2kr} + \beta^{2kr} + \alpha^{2r}\beta^{2r}\left(\alpha^{2(k-2)r} + \beta^{2(k-2)r}\right)$$
$$= L_{2kr} + L_{2(k-2)r}, \qquad \text{since} \quad \alpha\beta = -1.$$

5. Let $FI(a) = q$, $FI(b) = r$, $\text{lcm}(q, r) = \ell$.
Then $a \mid F_q \mid F_\ell$, $b \mid F_r \mid F_\ell$, so $\text{lcm}(a, b) \mid F_\ell$.
Now suppose $FI(\text{lcm}(a, b)) = m$. Then $m \mid \ell$.
On the other hand, $a \mid \text{lcm}(a, b)$, so $q \mid m$. Similarly, $r \mid m$, so $\ell \mid m$.
Thus $\ell = m$.

7. (1) We claim that $5F_n = L_{n+1} + L_{n-1}$.
    Set $n = 1$. Then $5 = L_2 + L_0 (= 3 + 2)$.
    Set $n = 2$. Then $5 = L_3 + L_1 (= 4 + 1)$.
(2) $F_{-1} = 1, L_{-1} = -1$.
    Set $n = 0$. Then $F_0 = 0 = -F_0$; and $L_0 = 2 = L_0$.
    Set $n = 1$. Then $F_{-1} = 1 = F_1$; and $L_{-1} = -1 = -L_1$.

$\boxed{\text{V}}$(3) As an example, let us look at the particular case

$$5F_n = L_{n+1} + L_{n-1}$$

If we know that, for some n,

$$5F_n = L_{n+1} + L_{n-1},$$

and

$$5F_{n+1} = L_{n+2} + L_n,$$

then, adding

$$5F_{n+2} = L_{n+3} + L_{n+1},$$

and, subtracting,

$$5F_{n-1} = L_n + L_{n-2}.$$

Thus we may proceed forwards or backwards indefinitely, to establish the identity for any value of n.

This argument works for any collection of sequences $\{u_n\}$ satisfying a (second order) linear recurrence relation $u_{n+2} = qu_{n+1} + pu_n$.

CHAPTER 4

1. A $\{\frac{b}{b-a}\}$-gon is just a $\{\frac{b}{a}\}$-gon described in the opposite sense. Since we cannot have $a = \frac{b}{2}$, because a, b are coprime, we must have $a < \frac{b}{2}$ or $a > \frac{b}{2}$. But $a > \frac{b}{2}$ if and only if $b - a < \frac{b}{2}$.

3. You can use the FAT algorithm; or you can just fold on any of the n crease lines systematically. (See Figure 8 for the case $n = 2$; 8(d) is, of course, the FAT pentagon.)

5. (1) The tape has (at the top or bottom) fold lines making angles

$$\frac{\pi a_i}{b}, \frac{2\pi a_i}{b}, \cdots, \frac{2^n \pi a_i}{b}$$

so long as these angles are acute, i.e., $\frac{2^n \pi a_i}{b} < \frac{\pi}{2}$ or $2^{n+1} a_i < b$. The biggest angle will be followed by an angle of the same size.

In item (9) we have every odd number $< \frac{1}{2} \times 31$ along the top, so we get folding instructions from the complete symbol for every $\{\frac{31}{a}\}$-gon with $a < \frac{1}{2} \times 31$. But this includes *all* star $\{\frac{31}{a}\}$-gons (see answer to BREAK 1). There is nothing special in this argument about 31 — any odd number would work, but, of course if b is not prime one must be careful to take a prime to b.

(2)
$$\begin{array}{c|cccc|ccccccc} 91 & 3 & 11 & 5 & 43 & 9 & 41 & 25 & 33 & 29 & 31 & 15 & 19 \\ & 3 & 4 & 1 & 4 & 1 & 1 & 1 & 1 & 1 & 2 & 2 & 3 \end{array}$$

(Only odd numbers prime to 91 appear in the top row.)

[V] (3) Answers will vary, of course.

(4)
$$\begin{array}{c|ccc} 33 & 9 & 3 & 15 \\ & 3 & 1 & 1 \end{array}, \text{ the bottom row is as in } \begin{array}{c|ccc} 11 & 3 & 1 & 5 \\ & 3 & 1 & 1 \end{array}. \text{ Thus}$$

$$\frac{33}{9} = \frac{11}{3}, \quad \frac{33}{3} = \frac{11}{1}, \quad \frac{33}{15} = \frac{11}{5},$$

the reduction factor 3 being the same in all cases.

(5)
$$\begin{array}{c|cc} 7 & 1 & 3 \\ & 1 & 2 \end{array}, \text{ so we expect (and get!) } \begin{array}{c|cc} 91 & 13 & 39 \\ & 1 & 2 \end{array}.$$

7. (1) It tells us that the quasi-order of 2 mod 91 is 12 and that, in fact,

$$2^{12} \equiv 1 \bmod 91$$

(2)
$$\begin{array}{c|cccccc} 23 & 1 & 11 & 3 & 5 & 9 & 7 \\ & 1 & 2 & 2 & 1 & 1 & 4 \end{array}, \text{ so } 2^{11} \equiv 1 \bmod 23; \text{ hence } 2^{11} - 1,$$

which is obviously bigger than 23, is not a prime, destroying Mersenne's hope. (On the other hand, the formula $2^p - 1$ will produce many primes.)

$$(3) \quad 641 \begin{array}{|ccccccccc|} 1 & 5 & 159 & 241 & 25 & 77 & 141 & 125 & 129 \\ 7 & 2 & 1 & 4 & 3 & 2 & 2 & 2 & 9 \end{array}, \text{ so}$$

$$2^{32} \equiv -1 \bmod 641$$

hence $2^{2^5} + 1$, which is obviously bigger than 641, is not prime.

9. (1) $\quad 13 \begin{array}{|ccc|} 6 & 5 & 2 \\ 1 & 1 & 1 \\ 1 & 1 & 1 \end{array} 4$

(2) $\quad 17 \begin{array}{|ccccccc|} 3 & 4 & 6 & 8 & 1 & 7 & 2 \\ 1 & 1 & 1 & 2 & 1 & 1 & 1 \\ 0 & 1 & 0 & 0 & 0 & 1 & 1 \end{array} 5$

CHAPTER 5

1. [V](1) You are on your own here.

   (2) You should find that you never need more than four colors.

3. [V](1) To show that the graph of Figure 1(b) is isomorphic to the graph of Figure 1(a), put a on top of 3, c on top of 4, b on top of 1, and d on top of 2.

   To show that the graph of Figure 1(c) is isomorphic to the graph of Figure 1(a) put v on top of 3, x on top of 4, u on top of 1, and w on top of 2.

   (2) Under the conditions given, the required graphs are:

   one vertex   •   one graph

   two vertices   • •   •——•   two graphs

   three vertices   four graphs

   (3) One obvious conjecture is that the number of graphs doubles as you add one more vertex. Hence the number of graphs on n vertices is 2^{n-1}. Unfortunately, this is not true. We show the *eleven* graphs on four vertices below. It turns out that there is a complicated formula for the number of graphs on n vertices that uses Pólya

enumeration techniques (see Section 4 of Chapter 8). This is somewhat difficult to use in practice.

no edges

one edge

two edges

three edges

four edges

five edges

six edges

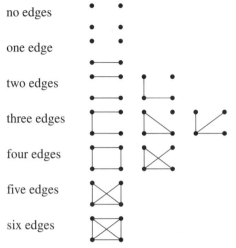

(The symmetry of this list should give you cause to pause.)

5. (1) Look at the degrees of the vertices. Look at the cycles.

(2) $K \not\cong L$. The graph L has two 3-cycles, whereas K has no 3-cycles.

7. Every vertex in K_n has degree $n - 1$. Hence the number of edges in K_n is $\frac{n(n-1)}{2}$. This comes about because there are n vertices (hence $n(n - 1)$), but in this count we have counted every edge twice (hence $\frac{1}{2}n(n - 1)$).

This result can be generalized. For any graph G,

$$\sum_{v \in V(G)} \deg v = 2E$$

where E is the number of edges of the graph G. This comes about in the same way. If we add all the degrees of G, we are adding all the edges, but each is counted twice. So $\sum_{v \in V(G)} \deg v = 2E$.

9. **Theorem** *Let G be a connected multigraph. Then there exists a route covering all of the edges and starting at u and ending at v if and only if the degrees of u and v are odd and all other vertices have even degree.*

Proof Suppose there is such a route starting at u and ending at v. Then Euler's argument, counting the number of edges as you go through a given vertex, shows that u and v have odd degrees and every other vertex has even degree. Suppose u and v have odd degrees and all the rest have even degrees. Add an edge to G linking u and v. Then the multigraph $G + uv$ satisfies Euler's Theorem from the text. Hence there is an Euler tour in $G + uv$. Removing uv from this tour gives the route we want for this theorem. □

11. (1) Let M be a map and $D(M)$ its dual. Assume that $D(M)$ is bipartite. Then every face of $D(M)$ has an even number of edges because every cycle in a bipartite graph is even. But every face of $D(M)$ corresponds to a vertex in M. Hence every vertex in M has even degree. Does the converse hold?

(2) Start at any vertex v and color it white. Now color all the neighbors N_v of v black. Next color all the neighbors of the vertices N_v white. Continue until all vertices are either white or black. Then you have your bipartition.

If at some stage you should try to color a white vertex black or a black vertex white, then this can only be because the graph has an odd cycle. But trees have no cycles. So you won't ever be trying to color a white vertex black, or vice versa.

13. (1) No. In Figure 6, K_4 is drawn with two edges crossing. But K_4 is planar, as the drawing of Figure 15(a) shows.

(2) No. The graph you get is shown below.

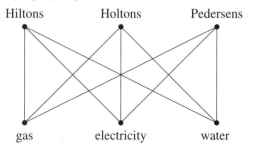

This graph is $K_{3,3}$.

We show in the next piece of text that this graph is non-planar.

15. See [6] or [8].

17. Suppose the Petersen graph is 3-edge colorable. Color its edges in red, white, and blue. Now look at the subgraph consisting of just the red and white edges. This will be regular of degree 2. So it must be an even cycle or a collection of even cycles. Now the Petersen graph has no 4-cycles. Hence the red–white graph *has* to be a 10-cycle. This is impossible, since the Petersen graph isn't Hamiltonian. So no red–white graph exists. So we can't 3-edge color the Petersen graph.

We show a 4-edge coloring below.

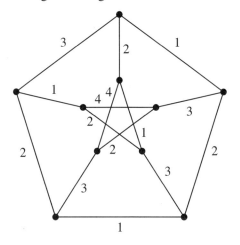

19. (1) This is only possible for even values of n. In that case the graph is a collection of disconnected edges.

(2) This is only possible for $n \geq 3$. These graphs are various combinations of cycles.

(3) This is definitely too hard! No one has been able to characterize all graphs that are regular of degree 3.

We can do the problem for $n = 4$ and $n = 6$, though. $\underline{n = 4}$. Here every vertex has to be joined to every other vertex. So $G = K_4$. $\underline{n = 6}$. The easiest way to do this is to first find $\overline{G}$, the complement[1] of G. Now $\overline{G}$ is regular of degree 2. Hence $\overline{G}$ is C_6 or two C_3's. If $\overline{G} = C_6$, then G is the graph below.

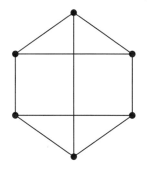

If $\overline{G} =$ two C_3's, then $G = K_{3,3}$.

---

[1] The complement of a graph G is the graph $\overline{G}$ with the same vertices as G and having an edge joining the vertices v and w precisely if G doesn̓t.

(4)

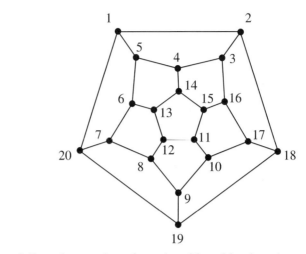

Just follow the numbers from 1 to 20 and back to 1.

(5) Since K_5 is a subgraph of all K_n for $n \geq 5$, we don't even have to sprout measles to use Kuratowski's Theorem.

$K_{1,n}$ is planar because it clearly doesn't contain K_5 or $K_{3,3}$. Similarly, $K_{2,n}$ is planar for $n \geq 2$. All other complete bipartite graphs contain $K_{3,3}$.

(6)

These, of course, are all the trees on 5 vertices.

$\boxed{\text{V}}$(7)

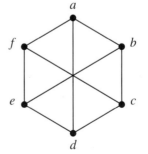

By Vizing's Theorem we know we need three or four colors. So suppose we can do it in three. We have to use all three at a, so color ab in red, ad in green, and af in black.

Suppose bc is colored in green and be in black. Then fc is colored in red, and so fe is colored in green. This forces us to color *de* in red and then *cd* in black. Hence we have a 3-edge coloring of $K_{3,3}$.

(8) If G is regular of degree 3, then $\sum_{v \in VG} \deg v = 2e$ gives us $3n = 2e$, where G has n vertices. But $2e$ is even. Hence n is even.

Now suppose G is Hamiltonian. If H is a Hamiltonian cycle in G, it has an even number of vertices and hence an even number of edges. Color the edges of H alternately 1 and 2.

Since G is regular of degree 3, then the edges in G but not in H are just a collection of independent edges. Color these edges 3.

We have now 3-edge colored G. Hence, if all planar graphs of degree 3 are Hamiltonian, the Four-Color Theorem must be true. (Clearly, there must be some non-Hamiltonian planar graphs of degree 3. Try to find one. You shouldn't bother with any graph with fewer than 38 vertices!)

CHAPTER 6

1. (1)

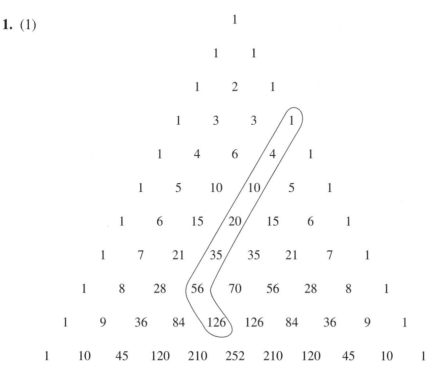

(2) If we subtract $\binom{n+k}{n \ k}$, the last summand on the left, from both sides, the assertion is converted into the equivalent assertion

$$\sum_{i=0}^{k-1} \binom{n+i}{i \ \ n} = \binom{n+k}{k-1 \ \ n+1}$$

This has exactly the same form as the original assertion, but k is replaced by $(k-1)$. Thus we may continue in this way, eventually reaching the equivalent assertion

$$\sum_{i=0}^{0} \binom{n+i}{i \ \ n} = \binom{n+1}{0 \ \ n+1}$$

or

$$\binom{n}{0 \ \ n} = \binom{n+1}{0 \ \ n+1}$$

But this is palpably true!

If we argue similarly to prove

$$\binom{n}{r-1 \ \ s+1} + \sum_{i=0}^{k} \binom{n+i}{r+i \ \ s} = \binom{n+k+1}{r+k \ \ s+1}$$

we keep subtracting until we reach

$$\binom{n}{r-1 \ \ s+1} + \binom{n}{r \ \ s} = \binom{n+1}{r \ \ s+1}$$

But this is just the Pascal Identity.

(3), (4), (5) are explained in the text.

(6) When parallelograms are mentioned below they will always refer to a parallelogram with two sides running parallel to the direction in which r is constant and the other two sides running parallel to the direction in which s is constant. If you think of Figure 7 as it sits within the Pascal Triangle, the following should be clear.

First calculate the sum of the entries inside, or on the boundary of, the parallelogram having the top vertex located at $\binom{0}{0 \ 0}$ and the bottom vertex located at $\binom{n+k+\ell}{r+k \ \ s+\ell}$. If we call this sum A, then, according to the result of part (5) above,

$$A = \binom{n+k+\ell+2}{r+k+1 \ \ s+\ell+1} - 1$$

Next, subtract the sum of all entries in the parallelogram having the top vertex located at $\begin{pmatrix} 0 \\ 0 & 0 \end{pmatrix}$ and the bottom vertex located at $\begin{pmatrix} n+k-1 \\ r+k & s-1 \end{pmatrix}$. If we call this sum B, then, by part (5),

$$B = \begin{pmatrix} n+k+1 \\ r+k+1 & s \end{pmatrix} - 1$$

Now subtract the sum of all entries in the parallelogram having the top vertex located at $\begin{pmatrix} 0 \\ 0 & 0 \end{pmatrix}$ and the bottom vertex located at $\begin{pmatrix} n+\ell-1 \\ r-1 & s+\ell \end{pmatrix}$. If we call this sum C, then, by part (5),

$$C = \begin{pmatrix} n+\ell+1 \\ r & s+\ell+1 \end{pmatrix} - 1$$

But we have now subtracted the entries in the parallelogram having the top vertex located at $\begin{pmatrix} 0 \\ 0 & 0 \end{pmatrix}$ and the bottom vertex located at $\begin{pmatrix} n-2 \\ r-1 & s-1 \end{pmatrix}$ *twice*. Hence we need to add those values back. Call their sum D; then

$$D = \begin{pmatrix} n \\ r & s \end{pmatrix} - 1$$

Now we see that the answer to our question is $A - B - C + D$, which simplifies to

$$A - B - C + D = \begin{pmatrix} n+k+\ell+2 \\ r+k+1 & s+\ell+1 \end{pmatrix} - \begin{pmatrix} n+k+1 \\ r+k+1 & s \end{pmatrix}$$
$$- \begin{pmatrix} n+\ell+1 \\ r & s+\ell+1 \end{pmatrix} + \begin{pmatrix} n \\ r & s \end{pmatrix}$$

3. (1) The number of ways we can put n students into 3 groups, having r, s, t members, respectively, so that the well-dressed student (WDS) goes into the first group, is $\begin{pmatrix} n-1 \\ r-1 & s & t \end{pmatrix}$; likewise, with WDS in the second group, the number is $\begin{pmatrix} n-1 \\ r & s-1 & t \end{pmatrix}$; and in the third group $\begin{pmatrix} n-1 \\ r & s & t-1 \end{pmatrix}$. The identity (25) now follows immediately.

(2) To show arithmetically that

$$\begin{pmatrix} n \\ r & s & t \end{pmatrix} = \begin{pmatrix} n-1 \\ r-1 & s & t \end{pmatrix} + \begin{pmatrix} n-1 \\ r & s-1 & t \end{pmatrix} + \begin{pmatrix} n-1 \\ r & s & t-1 \end{pmatrix}$$

or that

$$\frac{n!}{r!s!t!} = \frac{(n-1)!}{(r-1)!s!t!} + \frac{(n-1)!}{r!(s-1)!t!} + \frac{(n-1)!}{r!s!(t-1)!}$$

begin by using the *optimistic strategy*; namely, factor $\frac{(n-1)!}{r!s!t!}$ from each term on the RHS, obtaining the equivalent (conjectured) statement

$$\frac{n!}{r!s!t!} = \frac{(n-1)!}{r!s!t!}(r+s+t)$$

Since $r+s+t = n$, the last statement is true.

The algebraic proof imitates that for binomial coefficients: Write

$$(a+b+c)^n = (a+b+c)(a+b+c)^{n-1}$$

expand $(a+b+c)^n$ and $(a+b+c)^{n-1}$, and then compare the coefficients of $a^r b^s c^t$ on the left and right of the equality.

(3) They are just the numbers in the Pascal Triangle.

5. (1) Answer given in text.

(2) $W\left(P_{(k_2)}\right) = \dfrac{(k_2-a_1)!(k_2-a_2)!(k_2-a_3)!(k_2-a_1-a_2-a_3)!}{k_2!(k_2-a_1-a_2)!(k_2-a_1-a_3)!(k_2-a_2-a_3)!}$

$W\left(P_{(k_3)}\right) = \dfrac{(k_3-a_1)!(k_3-a_2)!(k_3-a_3)!(k_3-a_1-a_2-a_3)!}{k_3!(k_3-a_1-a_2)!(k_3-a_1-a_3)!(k_3-a_2-a_3)!}$

(3) Clearly, $W(P_{(k_2)})$, $W(P_{(k_3)})$ are deducible from $W(P_{(k_1)})$. Indeed, each of the three is deducible from $W(P^{(n)})$, using the "interchange of top and bottom" that we noticed in the simpler binomial case.

(4) Since we asked for a guess, your answer can't be wrong, but here is what these weights really turn out to be.

$$W\left(P^{(n)}\right) = \frac{n!\,\prod_{i<j}(n-a_i-a_j)!(n-a_1-a_2-a_3-a_4)!}{\prod_i(n-a_i)!\,\prod_{i<j<\ell}(n-a_i-a_j-a_\ell)!}$$

$$W\left(P_{(k_1)}\right) = \frac{\prod_i(k_1-a_i)!\,\prod_{i<j<\ell}(k_1-a_i-a_j-a_\ell)!}{k_1!\,\prod_{i<j}(k_1-a_i-a_j)!(k_1-a_1-a_2-a_3-a_4)!}$$

CHAPTER 7

1.

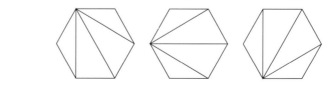

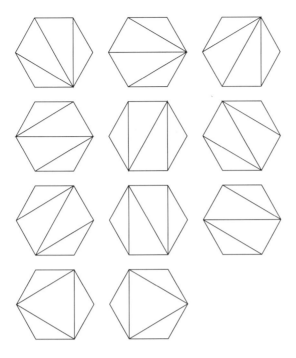

Thus we see that $c_4 = 14$.

3. (1) The trees below correspond to the parenthesized expressions of BREAK 2, part (1).

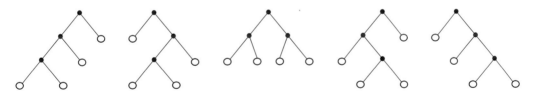

(2) The trees below correspond to the parenthesized expressions of BREAK 2, part (2), and the diagrams of BREAK 1.

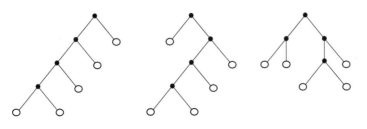

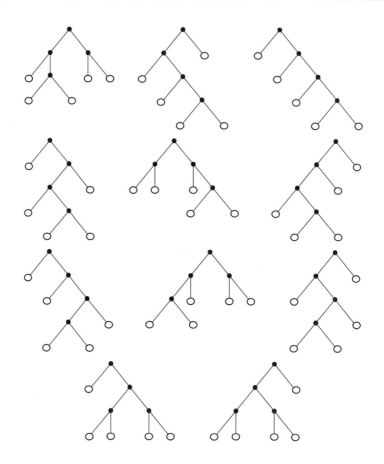

5. The answers to parts (1) and (2) are indicated in the text.

(3) $((s_1((s_2s_3)(s_4(s_5s_6))))s_7)$.

(4)

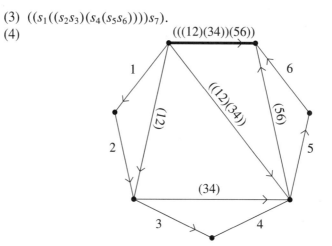

(5)

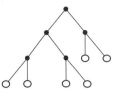

7. (1) No, suppose there were only 10 votes. Then, since A wins, the probability is 1.

(2) We count good paths from $(1, 0)$ to $(12, 5)$. We get

$$\binom{16}{5} - \binom{16}{4} = 4368 - 1820 = 2548.$$

We count all the paths from $(0, 0)$ to $(12, 5)$. We get $\binom{17}{5}$, so the probability of a good path is

$$\frac{2548 \times 5 \times 4 \times 3 \times 2}{17 \times 16 \times 15 \times 14 \times 13} = \frac{7}{17}.$$

If we replaced $(12, 5)$ by (a, b), we would get a proof of the Ballot Problem formula.

(3) $d_0 = 1, d_1 = 1, d_k = \frac{1}{k}\binom{2k}{k-1}$, so

$d_3 = \frac{1}{3}\binom{6}{2} = 5$

$d_4 = \frac{1}{4}\binom{8}{3} = 14$

$d_5 = \frac{1}{5}\binom{10}{4} = 42$

$d_6 = \frac{1}{6}\binom{12}{5} = 132$

$d_7 = \frac{1}{7}\binom{14}{6} = 429$

$d_8 = \frac{1}{8}\binom{16}{7} = 1430$

$d_9 = \frac{1}{9}\binom{18}{8} = 4862$

$d_{10} = \frac{1}{10}\binom{20}{9} = 16796$

(4) $c_0 = 1$

$c_1 = 1$

$c_3 = 5$

$c_4 = 14$

$c_5 = 42$

$c_6 = 132$

$c_7 = 429$

$$c_8 = 1430$$
$$c_9 = 4862$$
$$c_{10} = 16796$$

9. (1)
$$\frac{\left(-\frac{1}{2}\right)\left(-\frac{3}{2}\right)\cdots\left(-\frac{2r-1}{2}\right)}{r!} = (-1)^r \frac{1 \cdot 3 \cdots (2r-1)}{2^r r!}$$

$$= (-1)^r \frac{2r!}{2^{2r}(r!)^2}$$

$$= (-1)^r \binom{2r}{r} 2^{-2r}$$

(2)
$$\binom{n}{r} = (-1)^r \binom{r-n-1}{r}$$

$$\binom{n}{r-1} = (-1)^{r-1} \binom{r-n-2}{n-1}$$

so

$$\binom{n}{r} + \binom{n}{r-1} = (-1)^r \left\{ \binom{r-n-1}{r} - \binom{r-n-2}{r-1} \right\}$$

$$= (-1)^r \binom{r-n-2}{r}$$

$$= \binom{n+1}{r}$$

11. (1) $p = 4$, $q = 3$, $k = 4$, $n = 14$, $m = 3$, so, by (34), the required number is

$$d_{34}\binom{-2}{0} = d_{34} = \frac{1}{13}\binom{13}{3} = \frac{12 \cdot 11}{3 \cdot 2} = 22$$

(2) $n = pk - r$, so $\left[\frac{n}{p}\right] = k - 1$, and the required number is

$$d_{qk}\binom{-r}{0} = d_{qk}$$

A good path to $(k, (p-1)k-1)$ must finish by climbing vertically from $(k, (p-1)k - p)$.

CHAPTER 8

1. (1) Obviously,

$$\alpha_1 + \alpha_2 + \alpha_3 = -a_1$$
$$\alpha_1\alpha_2 + \alpha_1\alpha_3 + \alpha_2\alpha_3 = a_2$$
$$\alpha_1\alpha_2\alpha_3 = -a_3$$

Thus

$$\alpha_1^2 + \alpha_2^2 + \alpha_3^2 = (\alpha_1 + \alpha_2 + \alpha_3)^2 - 2(\alpha_1\alpha_2 + \alpha_1\alpha_3 + \alpha_2\alpha_3)$$
$$= a_1^2 - 2a_2$$
$$\alpha_1^3 + \alpha_2^3 + \alpha_3^3 - 3\alpha_1\alpha_2\alpha_3 = (\alpha_1 + \alpha_2 + \alpha_3)$$
$$\times (\alpha_1^2 + \alpha_2^2 + \alpha_3^2 - \alpha_1\alpha_2 - \alpha_1\alpha_3 - \alpha_2\alpha_3)$$

so

$$\alpha_1^3 + \alpha_2^3 + \alpha_3^3 = -3a_3 - a_1(a_1^2 - 3a_2) = -a_1^3 + 3a_1a_2 - 3a_3$$

(2) $\alpha_1^4 + \alpha_2^4 + \alpha_3^4 = (\alpha_1^2 + \alpha_2^2 + \alpha_3^2)^2 - 2(\alpha_1^2\alpha_2^2 + \alpha_1^2\alpha_3^2 + \alpha_2^2\alpha_3^2)$

$$= (a_1^2 - 2a_2)^2$$
$$- 2\left\{(\alpha_1\alpha_2 + \alpha_1\alpha_3 + \alpha_2\alpha_3)^2 - 2\alpha_1\alpha_2\alpha_3(\alpha_1 + \alpha_2 + \alpha_3)\right\}$$
$$= a_1^4 - 4a_1^2a_2 + 4a_2^2 - 2(a_2^2 - 2a_1a_3)$$
$$= a_1^4 - 4a_1^2a_2 + 4a_1a_3 + 2a_2^2$$

(3) $\frac{F_n}{F_m} = \frac{\alpha^n - \beta^n}{\alpha^m - \beta^m}$. If $m \mid n$, then the polynomial $\alpha^n - \beta^n$ is divisible by the polynomial $\alpha^m - \beta^m$, so $\frac{F_n}{F_m}$ is a symmetric polynomial in α, β with integer coefficients, and hence a polynomial in $(1, -1)$ with integer coefficients, so certainly an integer. (Recall that α, β are the roots of $x^2 - x - 1 = 0$, so that $\alpha + \beta = 1$, $\alpha\beta = -1$.)
$\frac{L_n}{L_m} = \frac{\alpha^n + \beta^n}{\alpha^m + \beta^m}$. If $m \mid n$ oddly, then, as above, $\frac{L_n}{L_m}$ is a symmetric polynomial in α, β with integer coefficients, and hence a polynomial in $(1, -1)$ with integer coefficients, so certainly an integer.

3. (1) The order of G is 12. The subgroups are:

G itself of order 12.
The remainders mod 6 represented by 0 2 4 6 8 10; the order is 6.
The remainders mod 4 represented by 0 3 6 9; the order is 4.
The remainders mod 3 represented by 0 4 8; the order is 3.
The remainders mod 2 represented by 0 6; the order is 2.
And, finally,
The remainder 0; the order is 1.

(2) The function $gH \longmapsto Hg^{-1}$ between the set of left cosets and the set of right cosets is well-defined. For if $g_1 \in gH$, then $g_1 = gh$ for some $h \in H$, so $g_1^{-1} = h^{-1}g^{-1} \in Hg^{-1}$. It is plainly a one-to-one correspondence between the set of left cosets and the set of right cosets.

5. (1)

$$(1\ 2\ 3\ 4\ 5) \longrightarrow \begin{cases} (1\ 2\ 3\ 4\ 5) & \text{(identity)} \\ (2\ 3\ 1\ 4\ 5) & \text{(rotation through } \frac{2\pi}{3} \text{ about axis 45)} \\ (3\ 1\ 2\ 4\ 5) & \text{(rotation through } \frac{4\pi}{3} \text{ about axis 45)} \\ (3\ 2\ 1\ 5\ 4) & \text{(interchanging the poles)} \\ (2\ 1\ 3\ 5\ 4) & \text{(interchange plus rotation through } \frac{2\pi}{3}) \\ (1\ 3\ 2\ 5\ 4) & \text{(interchange plus rotation through } \frac{4\pi}{3}) \end{cases}$$

(2)

Edge 12 has 3 homologues

Vertex 4 has 2 homologues

Face 124 has 6 homologues

7. $\boxed{V}$ (1) As in the example

$$\begin{pmatrix} 1 & 2 & 3 & 4 & 5 & 6 & 7 & 8 & 9 & 10 & 11 \\ 2 & 4 & 7 & 1 & 3 & 11 & 5 & 6 & 8 & 10 & 9 \end{pmatrix}$$

we start with 1 and continue till we get back to 1, as eventually we must. Thus

$$1 \to 2 \to 4 \to 1$$

giving us the cycle (1 2 4). We then start again with the lowest number not already reached, in this case 3, and repeat the procedure

$$3 \to 7 \to 5 \to 3$$

Thus we eventually exhaust *all* the numbers being permuted, and have expressed our permutation as a composition of cycles acting on mutually disjoint sets of numbers.

(2) Explained in the text.

(3) $(\iota_1\ \iota_2)(\iota_1\ \iota_3) \cdots (\iota_1\ \iota_r)$

9. (1) Obviously, Q satisfies the conditions for a subgroup. If P contains only even permutations, then $Q = P$, so the index of Q is 1. Otherwise, if ρ is an odd permutation in P, then P consists of the (disjoint) cosets Q and $Q\rho$, so the index is 2. The first case occurs if $P = A_n$, the second if $P = S_n$. (Of course, you could find other examples.)

(2) Let $g \in G$, $g \notin H$. Then G is the disjoint union $H \cup gH$; and it is also the disjoint union $H \cup Hg$. Hence $gH = Hg$.

CHAPTER 9

1. We get around to this later in the section.

3. See BREAK 2. If $n < s$, then we are unable to get an s-anticlique.

5. No. The argument is rather a case-by-case analysis, so we'll only start it off. From the argument of the text we have two cases. We assume that there is only *one* monochromatic triangle. It must arise in one of the two following ways.

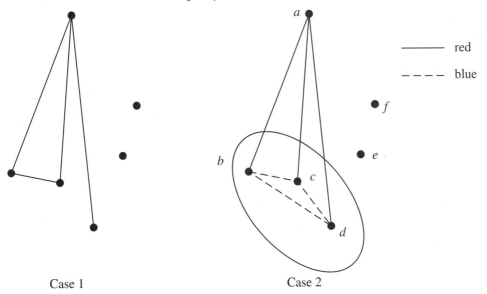

Case 1   Case 2

We will discuss Case 2. Then at most one of the edges from f to $\{b, c, d\}$ is blue. Hence at least two of these edges are red. Hence af is blue. Similarly, ae is blue.

This forces ef to be red. But then we have a red triangle involving e, f, and one of the vertices of $\{b, c, d\}$.

Now you should try to argue Case 1.

7. $\boxed{V}$ Look at the argument of the next piece of text based on Figure 3. So $N(3, 4) > 8$. (The fact that $N(3, 4) = 9$ follows in the text.)

9.

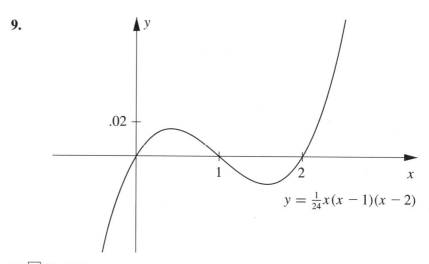

$$y = \tfrac{1}{24}x(x-1)(x-2)$$

11. $\boxed{\text{V}}$ (1) This graph has a matching that saturates G, for example, $(g_1 b_1, g_2 b_3, g_3 b_4)$. You should have checked that, for any $S \subseteq G$, $|S| \leq |N(S)|$. For instance, if $S = \{g_1, g_2\}$, $N(S) = \{b_1, b_2, b_3, b_4, b_5\}$, and the required inequality holds.

(2) Let $S = \{g_3, g_4, g_5\}$. Then $N(S) = \{b_4, b_5\}$. So $|S| > |N(S)|$. Clearly, the vertices of S cannot all be matched.

13. The number of edges leaving X is $r|X|$ and the number leaving Y is $r|Y|$. Hence $r|X| = r|Y|$, which shows that $|X| = |Y|$.

15. (1) $N(2, 2, 2) = 2$. As soon as you have an edge it must get a color.

$N(2, 2, 3) = 3$. If there is no monochromatic triangle in K_3 in a given color, one of the edges must be colored in one of the other colors.

$N(2, 3, 3) = 6$. If in K_6 there is no edge of the first color, there must be a monochromatic triangle in one of the other colors. For $n \leq 5$, K_n can be colored with two colors to avoid a monochromatic triangle.

(2) By the argument that we used to show that

$$N(3, 4) \leq N(2, 4) + N(3, 3)$$

we can show that $N(m, n) \leq N(m - 1, n) + N(m, n - 1)$. We now proceed by induction on $m + n$.

Now

$$N(2, k) = k \leq \binom{k}{1} \qquad \text{for all} \quad k \geq 2$$

Similarly,

$$N(k, 2) = k \leq \binom{k}{1} \qquad \text{for all} \quad k \geq 2$$

So assume that the result is true for all

$$\binom{n}{r \ \ s} \qquad \text{with} \quad r + s < m + n$$

Now $N(m, n) \leq N(m - 1, n) + N(m, n - 1)$.

Then

$$N(m, n) \leq \binom{m + n - 3}{m - 2} + \binom{m + n - 3}{m - 1}, \qquad \text{by the inductive hypothesis,}$$

$$= \binom{m + n - 2}{m - 1} \qquad \text{by the Pascal Identity (see Chapter 6).}$$

This establishes the inequality.

(3) Suppose $G - \{v\}$ has a component, C, that has one edge linking it with v. Then in $G - \{v\}$ this produces a single vertex of degree 3 in C. This is not possible, because every graph must have an even number of vertices of odd degree.

Hence every component of $G - \{v\}$ must be joined to v by 2 or 4 edges, so that G is as in Figure 1 (below).

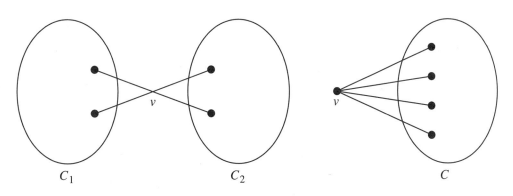

FIGURE 1

In the graph on the left $|C_1| + |C_2|$ is odd, since G has an even number of vertices and v is the only vertex in neither C_1 nor C_2. Hence at most one of C_1, C_2 is an odd component. In the graph on the right, C is the only odd component. A graph based on Figure 2 (on the next page) will not have perfect matching.

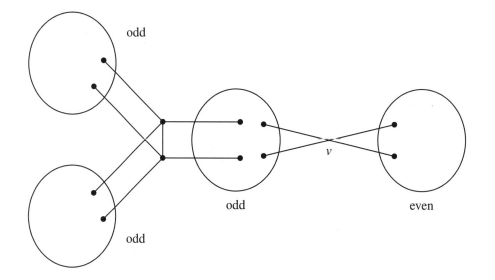

FIGURE 2

Index

10-gon, 83
124-year history, 127
2-colorable, 140
2-factors, 296
2-period, 83
2-symbol, 90
2-value, 53
2×2 table, 11
20-gon, 83
3-edge colorable, 153
3-edge coloring, 151
3-good paths, 224 (Fig), 232 (Fig)
3-period folding, 88
31-gon, 90
4-dimensional space, 172
4-face coloring, 151
5-colorable, 148
5-gon, 83 (Fig)
7-gon, 80 (Fig), 244
8-gon, 76 (Fig)
9-gon, 82
(2, 1)-folding procedure, 78
(2, 1)-tape, 78
(m, n), 83
$(p + 1)$-gons, 215

A is a symmetric figure, 251
A is more symmetric than B, 248

A_4, 263
   alternating subgroup of S_4, 267
A_5, 253
A_n, 248, 266
Abstract, 4, 14, 18
Accuracy, 3
Adding
   fractions, 9
   machine tape, 74
   rational numbers, 9
Advanced versions, vii
Aldred, Robert, 198
Alexanderson, G. L., 190, 198
Alexandria, 26
Algebra, 237
   of polynomials, 236
Algebraic, 162
Algorithm, 81, 95, 114
Allowed transformations, 237
Alternating
   form (in n variables), 263
   group, 263
   group A_4, 248
   group on n objects, 266
   signs, 229
Ambient
   3-dimensional space, 246
   4-dimensional space, 246
   plane, 244
American Mathematical Monthly, 10

Anchor, 172
Ancient Greeks, 73
André, Désiré, 210, 212, 233
André Reflection Method, 210 (Fig), 217
Annals of Mathematics, 45
Answers, ix
   to breaks, 299ff
Anticliques, 271
Aphorism, 2
Appel, K., 127, 149, 157
Appetizers, viii
April 1, 44
Arctic pentagon, 114
Arithmetica, 26, 28
Arithmetical interpretation, 162
Associative law, 239
At random, 4
Athenaeum, 130
Attractions of mathematics, viii
Averages, 8
Awakening, 227
Axiom, 2
Axiomatic systems, 15

Babylonians, 29
Ballot problem, 212, 213, 317
Base, 99
Baseball, 3, 8
Batting averages, 3
Beckenbach, Edwin F., 20
Beineke, L. W., 157
Bekes, Robert, x
Bell, E. T., 40
Bereiter, Carl, 269
Berge, C., 238
Bertrand, Joseph Louis François, 212
Bibliography, 200
Biggs, N. L., 158
Binary operation, 201
Binet formulae, 52, 237
Binomial
   coefficient, 159, 209
   theorem, 218
Bipartite, 289
   graph, 141 (Fig)
Bipartition, 308
Birthdays, 285
Bisect, 87
Bisecting angles, 73
Blades, 169, 170
   northeast, 170
   northwest, 170

Blanuša, Danila, 156
Bobby pin, 96, 109
Books, 96
Borcherd, Richard, 235
Borgia, 12
Borromean rings, 107 (Fig)
Bosses, 276
Boundaries, 194
Bracelet, 109
Braid, 149
Braided Platonic solids, 249
BREAKS, ix
Breuil, Christophe, 44
Bridge (of connected graph), 296
Bright and curious reader, viii
Building blocks, 235
Bürmann–Lagrange inversion formula, 217

Cambridge University, 288
Can
   of baked beans, 164
   of spaghetti, 164
Cardinality, 299
Carroll, Lewis, 127, 154
Cartan, Henri, 159, 236
Catalan, Eugene Charles, 199
Catalan numbers, 199ff
   calculation of, 217, 223
   classical, 199ff
   generalized, 229
   generalized, calculation of, 220
Children's Television Workshop, 16
Chinese, 29
Christians, 27
Christmas Stocking Theorem, 164
Circuit, 132, 203
Clark, J., 158
Classification of finite simple groups, 42
Clay Mathematics Institute, 285
Clifton College, 130
Clockwise, 246
Clothes, 97
C_n, 248
Coates, John, 41
Coconuts, 17
Coefficient, 162
Coincidences, 285
Collaborators, 276
Color, 258
Colored
   paper, 96
   side, 97

Coloring, 258, 262, 273

Coloring edges, 149

Combinatorial
  interpretation, 161, 170
  set theory, 275

Combinatorics, 238, 254

Combining fractions, 9

Common index, viii

Compact bounded configuration, 247

Comparative adjective, 6

Complete symbol, 91, 305

Complex number, 37

Components, 132

Composition of cycles, 264

Computer, 14, 40, 281ff

Conceptual
  model, 15
  thinking, 2

Concrete 4, 14

Concrete geometrical model, 236

Connected multigraph, 307

Conrad, Brian, 44

Construction
  paper, 96
  Euclidean, 73
  exact, 74
  explicit, 74

Contracted, 90, 114

Contradiction, 2

Conventional
  reasoning, 2
  usage, 5

Converge, 81

Convergence, 79, 86

Convergent infinite series, 168

Convex
  8-gon, 75
  hexagon, 205
  polygon, 199

Coordinate geometry, 171

Correct strategy, 20

Cosets,
  left, 320
  right, 320

Counter-intuitive, 19

Counterclockwise, 246

Counties of England, 128, 129 (Fig)

Counting p-good paths, 223

Complement (of a graph), 309

Coxeter, H. S. M., 74, 124

Cox, David, 23, 46

Crazy
  dice, 6
  ideas, 149

Crease lines, 75

Cross-product, 172

Cube, 99, 105, 106 (Fig) 107, 248, 266, 267 (Fig)

Cube, interior diagonal of, 261

Cubic identity, 66

Cuboctahedron, 195

Cusp form, 42

Cycle, 132, 240

Cycle
  index, 257ff
  of length 2, 240

Cyclic
  group of order n, 248
  permutation, 185 (Fig)

D^2U^1
  -folding procedure, 78
  -tape, 78

D_4, 258

D^mU^n, 83

D_n, 248

Dance, 287

Darmon, Henri, 44, 46

de Morgan, Augustus, 128

Degree (of a vertex), 132

Demystification, 228

Density, 299

Descartes, Blanche, 238

Diagonal cube, 108 (Fig), 248, 249 (Fig)

Diagonals, non-intersecting, 199

Diamond, Fred, 44

Dickson, L. Eugene, 39, 46

Difference, 9

Dihedral
  group, 248
  group D_5, 255

Diophantine
  equations, 27
  problems, 27

Diophantus, 26, 302

Dirichlet, Peter Gustav Lejeune, 37, 46

Discrepancy, 86

Dissected
  hexagon, 206
  polygon, 208, 216 (Fig)

Distance, 171, 172

Divisibility, 3, 52

Dodecahedron, 99, 108, 109 (Fig)ff, 157

Dodecahedron, Golden, 113, 246, 247 (Fig), 262

Don't believe everything, 1
Dot product, 155
DOWN TWICE, 77
Drift, 12
Dual graph, 139 (Fig)

Economics, 235
Edge
  -coloring, 150
  (of a graph), 130
Edmonds–Gallai Theorem, 296
Education, 2
Efficient matching, 295
Election, 213
Elliptic
  curves, 41
  curves, semi-stable, 44
Empty, 264
End node, 203, 206
Epsilon, 276
Equal, 4
Equality, 4, 222
Equilateral triangles, 176
Equilibrium, 235
Equivalence relation, 5
Erdős number, 276
Erdős, Paul, 275, 281, 288, 297
Erlanger Programm, 237
Error-correction, 88
Euclid, 26
Euclidean, 245
  algorithm, 68
  geometry, 237, 246
  invariant, 246
  motion, 244
  motion in 3-dimensional space, 246
  motions that reverse orientation, 244
  tools, 73
Euler, Leonhard, 36, 37, 46, 199, 307
Euler
  tour, 137
  tour problem, 136
Even
  and odd permutations, 263ff
  period, 89
  permutations, 248, 267
  vertices, 182
Everswear, 12
Exact sciences, 235
Exceptional case, 169
Existence
  of two-sided identity, 239

of two-sided inverses, 239
Expectation, 20
Experiment, 11
Exponents, 81
Extending the binomial coefficients, 218
Extensions, 194

Fabrice, Valognes, 20
Fair coin, 281
False proof, 130
Fantasy, 12, 227
FAT
  -algorithm, 74, 79, 305
  31-gon, 90
  7-gon, 79
Favorable outcome, 11
Fellow of the Royal Society (F.R.S.), 148
Fermat, 23ff, 46, 115
Fermat's Last Theorem (FLT), viii, 23, 28, 32
Fermat
  number, 82
  prime, 73
Fibonacci
  index, 54, 63
  numbers, 49ff
  sequence, 67
Fibonacci Quarterly, 69
Fibonaccian numbers, 69
Field, 151
Fields
  Medal, 45, 235
  John Charles, 45
Finite group, 248
Finite symmetry group, 251
Fisher, David, 20
Fixed point, 18
FLT, viii, 23, 39
Fold, 74
Fold
  down, 82
  lines, 97
  up, 82
Folding
  instructions, 305
  procedures, 72ff
Formulae, 14, 201
Four-color, 127
  conjecture, 130
  problem, 127
  theorem, 33, 127ff
Fraction, 19, 85, 240

Framework
    algebraic, 159
    combinatorial, 159
    geometric, 159
French Academy, 37
Frey, Gerhard, 43
Friedlander, Richard J., 20
Fritsch, G., 158
Fritsch, R., 158
Fundamental parallelepiped, 182, 183 (Fig)

G-geometry, 245
G-invariants, 245
Game, 5
Gardner, Martin, 154, 275
Gauss, Carl Friedrich 73
Gaussian polynomial, 192
gcd, 53, 68
General
    base *t*, 73
    quasi-order theorem, 114, 119, 122, 123
Generalization, 73, 115
Generalized
    binomial coefficient, 218
    Catalan numbers, 229
    complex number, 128
Generating function, 202
Generator, 248
Geometric, 173
    context, 238
    practice, 71
Geometry, 236, 239, 244, 262
    of a 3-dimensional analogue, 193
    plane Euclidean, 244
Germain, Sophie, 37, 46
Glue, 96
Gödel, Kurt, 2
Golden Dodecahedron, 113, 246, 247 (Fig), 262
Good paths, number of, 227
Göttingen, 39
Gould, H. W., 191, 198, 200, 233
Gouvêa, Fernando Q., 23, 36
Graph, 130ff, 203, 272ff, 323
Graph
    bipartite, 141 (Fig)
    complete 135
    complete bipartite, 141
    connected, 132 (Fig)
    disconnected, 132 (Fig)
    non-planar, 148
    planar, 144
    regular, 134

-theoretical, 272
theory, 136, 295
underlying, 149
Greeks, 27
Group, 239, 244, 246
Group,
    commutative, 241
    D_4, 258
    finite, 241
    Γ, 254
    infinite, 241
    of motions, 246
    of permutations, 258
    of permutations of *n* symbols, 248
    of symmetries, 259
    table, 241
    theorist, 290
    theory, 237
Groups acting on sets, 238
Gummed
    mailing tape, 96
    tape, 74
Guthrie, Frank, 128
Guthrie, Frederick, 127

Haken, W., 127, 149, 157
Hall, Philip, 288, 290, 297
Hall's Theorem, 292
Hamiltonian, 157, 308
Hamilton, William Rowan, 128
Handbook of Applicable Mathematics, 238
Hand
    calculators, 14
    towel, 96
Handles, 143 (Fig)
Hands-on-activities, 95
Harary, Frank, 238
Heawood, Percy John, 148
Hexagon, 177
Hexagonal cross-section, 197 (Fig)
Hexagon, regular, 180
Hexahedron, 106 (Fig), 266, 267 (Fig)
Hilbert, David, 285, 297
Hilton, Peter, 20, 46, 69, 124, 125, 198, 233, 269, 297
Hip roof, 112
History of the mathematical sciences, viii
Hoffman, Paul, 297
Hoggatt-Alexanderson Theorem, 191
Hoggatt, V. E., Jr., 190, 198
Holdsworth, John, x

Holton, Derek, 20, 46, 69, 124, 158, 198, 233, 269, 297
Homogeneous, 260
Homologue, 237, 254ff, 320
Human
 error, 79
 skill, 79
Hunting of the Snark, 127
Huzita, H., 125
Hyperplane, 172
Hyperstar of David Theorem, 184

Icosahedron, 99, 102 (Fig)ff
Icosahedron with symmetry group A_5, 253
Identity, 4
Identity element, 241
Imagination, 3
Independent
 edges, 289
 points, 247
Index, 244, 255
Index 2, 248
Inductive argument, 230
Industrial applications, 294
Inequalities, 194
Infinite sets, 299
Initial
 conditions, 228, 232
 configuration, 94
 integral lattice point, 211
Integer, 240
 coefficients, 260
Integral lattice, 208
 in the plane, 208
International Congress of Mathematicians, 45, 235
International Mathematical Union silver plaque, 45
Interpretation
 algebraic, 177
 combinatorial, 177
Interrelating geometry and algebra, 187
Intrigue, viii
Invariant, 246
Invariants, 245
Inverse, 241
 g^{-1}, 242
Isaac Newton Institute, 44
Isaacs, R., 155, 158
Isomorphic (graph), 131

Joe, 276

Jonah
 David, 221
 formula, generalized, 221
Jordan, Michael, 285
Justice, Dave, 10

k-colorable, 140
k-factors, 296
Kainen, P. C., 158
Kaliningrad, 136
Kempe, Alfred Bray, 33, 130, 148
Key formula, 59, 221
King Charles II, 148
Klein, Felix, 237, 244
Königsberg, 136
 bridge problem, 36, 136 (Fig)
Kummer, Ernst, 37, 39, 46
Kuratowski, Kazimierz, 146
Kuratowski's Theorem, 146, 310

Label, 206, 207 (Fig)
Lagrange, Joseph Louis, 242
Lagrange's Theorem, 243
Lamé, Gabriel, 37, 39, 46
Larcombe, P.J., 233
Learn, viii
Least common multiple (lcm), 62
Ledermann, Walter, 269
Legendre, Adrien Marie, 37, 46
Lengths, 182
LHS, ix
Lindemann, Ina, x
Linear, 3
Linear
 algebra, 151
 identity, 64, 67
 recurrence relation, 19
 relation, 65
Lively mathematical topic, vii
Lloyd, E.K., 158, 238
Logical thinking, 3
Logically superfluous, 211
Long fold lines, 113
Lovász, László, 297
Lucasian, 54, 62
Lucasian Numbers, 49ff, 69
Lucas Index, 54, 62, 63

Mackerelroe, 12
Male barber, 1
Map of England, 127, 129 (Fig)
Matching, 289, 322

Mathematical, 15
 abstraction, 15
 context, 19
 essays, viii
 model, 4
 reasoning, 2, 4
 relation, 19
 theory, 71
 usage, 5
Mathematical Reflections, vii
Mathematicians, vii, 295
Mathematics, 14, 235, 236
Mazur, Barry, 23, 47
Measles, 146ff
Memory, ix
Mersenne (Abbé), 115, 305
 number, 115
Metaphor, 258
Method of infinite descent, 34
Microcomputers, 14
Misunderstanding, 14
Miyaoke, Yoichi, 44
Models, 96
Modular function, 41
Monkeys, 14, 17
Monochromatic, 273
 cycle, 295
 triangle, 322
Mordell, Louis, 28
More
 likely, 5
 prosaic explanation, 228
Moslems, 27
Mountain folds, 97
MR, vii
Multi-dimensional, 3
Multigraph, 136
Multinomial coefficients, 170, 181
Multiplicative (notation), 242

Nash, John, 235
National Health Service, 2
Nearest neighbor, 190
Negative
 integer, 167, 219
 number, 18
Neighborhood
 (of a set S of vertices), 289
 (of a vertex), 289
Net diagram, 196
New
 century, 285
 identity, 184

Nobel Prize, 45, 235
Node, source, 206
Non
 -constructibility, 82
 -lattice point, 217
 -zero blades, 170
 -zero sextants, 169
North Pole, 113
Notable event, viii
Not treated, 11
Notational innovation, 159
Number
 -crunching, 295
 -theoretical, 26
 theory, 71
 tricks, 49
Numbers, 4
Numerical answers, 14
$N(r, s)$, 275

Oberkfell, Ken, 10
Octahedron, 99, 100, 101 (Fig)ff
Octahedron with symmetry group S4, 253
Odd
 degree (of vertex of graph), 280
 Lucasian numbers, 56
 period, 89
 permutations, 263
 vertices, 182
Oddly, 53, 237
One-to-one correspondence, 4, 204, 209, 243
Operation
 binary, 215
 ternary, 215
 p-ary, 215
Optimistic strategy, 77, 79, 87
Ordered
 pair, 5
 triple, 6
Ordering, 3
Order (of a group) 241
Orientation, 246
 -preserving symmetry of a polygon, 266
 reversing, 245 (Fig)
Over 100 years, 128

Paper
 -folding, 71
 clip, 105
 folders, 77
Paradox, 1ff
Parallelepiped, 171

Parrallelogram, 172
  fundamental, 172 (Fig), 183 (Fig)
Parenthesis, 201, 211
Parenthesized
  expression, 204 (Fig), 207 (Fig), 208, 211
  symbol, 206
Parenthesizing, 201
Parity, 31, 264
Partial, 3
Particular but not special case, ix, 206, 292
Parties, 271ff
Partitioning, 232
Partitioning
  good paths, set of, 225
  of path, 221 (Fig)
Parts (of vertex set), 289
Pascal Cuboctahedron, 196 (Fig)ff
Pascal
  Flower, 175 (Fig)
  Hexagon, 163 (Fig), 167, 169
  Identity, 160, 169, 170, 180, 220, 230, 323
  m-simplex, 188, 189
  Pyramid, 181, 197
  Tetrahedron, 170, 177 (Fig)ff
  Triangle, 159 (Fig)ff
  Windmill, 170
Path, 208, 228
  bad, 209, 231
  good, 209, 210, 212, 231
  p-good, 215, 230 (Fig)
Pattern, 259
  pieces, 96, 250 (Fig)
Pedersen, Jean, ix, 20, 46, 69, 124, 125, 198,
    233, 269, 297
Pedersen, Kent, x
Penrose, Roger, 41, 47
Pentagonal dipyramid, 97 (Fig)ff, 255 (Fig)ff
Penultimate stop, 229
Perfect
  in the mind, 79
  matching, 293, 323
Period, 81
  1, 81
  2, 83
Permutation, 92, 117, 238, 240, 257
  cyclic, 257, 264
  even, 238, 263, 320,
  exchanging a_1 and a_2, 186 (Fig)
  odd, 238, 263, 320
Petersen graph, 146, 154, 156, 308
Petersen, Julius Peter Christian, 154
Petersen Theorem, 296

Φ-algorithm, 116
Philanthropist, 20
Ph.D., 41
Physical context, 19
Pierce, Larry II, 181
Pigeonhole principle, 274, 284
Ping-Pong, 12
Planarity, 144
Plane $\mathbb{R}^2$, 237
Platonic solids, 99, 250 (Fig), 251 (Fig), 254, 267
Plouffe, Simon, 233
Plummer, Michael D., 297
Pólya
  Enumeration Theorem, 238, 254, 257
  George, 198, 238, 254, 263, 269, 306
  's personal notebook, 267
Polygon, 247
Polygonal dissection, 216 (Fig)
Polyhedra, 71, 73, 95
Polyhedral
  formula, 36
  symmetry, 239
Polyhedron, 247
Polynomial
  equations, 236
  identities, 52, 64
Polynomials
  identically equal, 22
  with rational coefficients, 219, 222
Pop-up model, 197
Popular
  imagination, 127
  misconception, 2
Positive
  integers, 3
  sector, 170
Power
  series, 202
  series expansion, 167
  series, formal, 202
Precise definition, 263
Preferred side, 206
Pregel, 136
Primary
  crease lines, 78
  folding procedure of period 3, 88
Primitive, 30
Princeton University, 44
Principle of Licensed Sloppiness, 16
Probabilistic
  argument, 281
  method, 277

Probability, 5, 7, 213, 286, 299
Problems, 19
Proof, traditional, 149
Prowess, 9
Proximand, ith, 94
Prussia, 136
Pseudo-partitioning, 232
Ψ-algorithm, 116
PT, 30
Putative angle, 77, 81, 82
Pythagoras, 28ff
Pythagoras' Theorem, 29
Pythagorean triple, 30, 300

q-analogue, 192
Quadratic identities, 57, 65ff
Quasi-order, 114
Quasi-Order Theorem, 72, 115
Quasi-regular
    $\{\frac{b}{a}\}$-gon, 74, 75
    N-gon, 72, 74
    pentagon, 82
    polygon, 71
    polygons, 93
    $\{\frac{7}{2}\}$-gon, 79
Quaternion, 128

Ramsey number, 275, 288, 295
Ramsey, F. P., 275, 283, 297
Ramsey Theory, 295
Rate, 16
Ratio, 16
Rational
    coefficients, 219
    number, 16, 240
Read, R. C., 238
Reader's fancy, viii
Reading proofs, ix
Real world, 4, 18
    constructions, 79
    problems, 14
    situation, 15
Reasoning mathematically, ix
Rectangular box, 172, 182
Recurrence
    formula, 232
    relation, 19, 202, 217, 229
Redfield, J. Howard, 238
Redfield–Pólya Theorem, 238
Reduced, 90, 93, 114

Reflection, 237, 244, 246
    method, 212
    procedure, 209
Regions, 139
Regions with different colors, 128
Regular
    convex 7-gon, 77
    convex n-gon, 71
    convex polyhedron, 99
    dodecahedron, 248
    hexagon, 206
    icosahedron, 248
    n-gon, 251
    octahedron, 248
    primes, 37
    star $\{\frac{b}{a}\}$-gon, 71
    tetrahedron, 194, 263
Relative notion, 247
Relevance, 19
Representative, 243
Reverse
    algorithm, 116
    symbol, 120
Rhombic dodecahedron, 180
Rhombic dodecahedron, regular, 180
RHS, ix
Ribenboim, Paulo, 23, 47
Ribet, Kenneth, 43
Right coset, 255
Robertson, N., 149, 158
"Roof"
    (of a hexagon), 161
    (of rhombic dodecahedron), 180
Rosen, Michael, 24
Rosenthal, William, 198
Ross, Peter, x, 10
Rotation, 237, 244
r-regular bipartite graph, 292
Rubber band, 105
Rubinstein, Joseph H., 269
Rule of thumb, 86
Russell, Bertrand, 1, 15

S_3, 242, 247
S_4, 248, 262
Saari, Donald G., 20
Saaty, T. L., 158
Sailors, 14, 17
Sam, 276
Sampling theory, 213
Sanders, D. L., 149

Satellites, 191

Saturate, 289, 322

Scalar, 3

Schlegel diagram, 157

Schoolboy invention, 127

Schwenk, Allen J., 20

Scientific American, 154

Scientific
    hypothesis, 11
    model, 15, 17

Scimemi, B., 125

Scioscia, Mike, 10

Scissors, 96

Scribbles in margins, 32

Scott, Richard, x

Secondary
    crease line, 83
    fold lines, 73

Self-evident truths, 2

Semi-regular hexagon, 176 (Fig)

Separable functions, 192

Serre, Jean-Pierre, 43

Set
    of colors, 259
    theory, 1, 4

Seymour, P. D., 149

Shallow bowl, 96

Shaw, George Bernard, 2

Sheehan, J., 158

Shimura, Goro, 41

Shimura–Taniyama Conjecture, 42ff, 46

Short lines, 108

Significant dates for FLT, 46

Simpson, E. H., 21

Simpson's Paradox, 9

Singh, Simon, 23, 47

Skepticism, 2

Skill, 3

Slaves, 276

Slide, 223

Sliding procedure, 224

Sloane, N. J. A., 233

S_n, 248

Snarks, 154

Somer, Larry, 69

Sophomore student, 181

Source node, 203

South Pole, 114

Speed of execution, 3

Sponge, 96

Sporadic finite groups, 235

Square One, 16

Stanley, Richard P., 233

Star
    b-gon, 72
    of David, 175 (Fig)
    of David Theorem, 173
    of David Theorem, original case, 191
    of David Theorem, generalized, 170, 171, 177, 181

Starred (★) material, ix

Statistical significance, 11

Statistical test, 11

Statistics, 11, 14

Stewart, Ian, 21

Strings (super), 235

Strip of paper, with parallel edges, 74

Subdivided hexagon, 206

Subgraph, 131, 147, 272

Subgroup, 242, 255, 319

Subgroup
    of index 2, 266
    of S_n, 266
    normal, 166

Subscript, preceding, 215

Supersymmetry, 235

Support, ith, 94

Supreme Facist, 276

Surprising features, 214

Symbol, 72, 90, 114

Symbol-manipulation, 18

Symmetric, 160, 236, 260

Symmetric
    function, 236
    group on n symbols, 248
    group S_3, 171, 184
    group S_n, 171
    polyhedron, 194
    polynomial, 236, 319

Symmetries
    of an equilateral triangle, 242
    of the square, 258

Symmetry, 159, 170, 209, 235ff, 239, 246, 254

Symmetry
    concept, 236
    group, 171, 237, 242, 255, 263
    group of A, 246
    group on 3 symbols, 247
    groups, 254
    groups of polyhedra, 248
    Identity, 159, 169, 180
    in geometry, 237, 239
    of a polyhedron, 266

theory, 235
Szekeres, George, 275, 288, 297

Tab, 196
Tait, P. G., 149
Taniyama, Yutaka, 41
Taylor, Richard, 44
Taylor series, 168
Telles, Roberta, x
Term-by-term differentiation, 168
Tetrahedron, 99, 100 (Fig) 266, 267 (Fig)
Tetranomial coefficients, 187
The Hunting of the Snark, 127,
Theoretical physicists, 235
Think, viii
Thinking, ix
Thomas, R., 149
Tombstone symbol (□), ix
Total, 3
Totient function, 36
Towers, 164ff
Traditional education, 3
Transitive, 3
Transitive relation, 5
Translation, 237, 244
Transposition, 264, 265
Treated, 11
Tree, 132, 203, 315 (Fig)
    2-ary, 214 (Fig)
    3-ary, 214 (Fig)
    binary, 203, 204 (Fig), 214 (Fig)
    diagram, 5, 7
    p-ary, 215
    rooted, 203, 206
    ternary, 214 (Fig)
Triangular
    dipyramid, 98, 256 (Fig)
    layers, 177
Tricolored faces, 111
Trihedral regions, 195
Trimming, 97
Trinomial
    coefficients, 159, 170
    expansion, 170
Trivalent map, 150 (Fig)
Truncated trihedral regions, 195
t-symbol, 90, 119
Tutte's
    Conjecture, 156
    Theorem, 295
Tutte, W. T., 156
Twist, 74

Type (of permutation), 257

Uncreased, 113
Unfavorable outcome, 11
Unified field theory, 235
Unique
    contracted symbol, 92
    factorization, 37, 38
Unit, 17
Universe, 235
Unlicensed sloppiness, 59
Unreliability of tests, 3
Unusual characteristics, 73
UP ONCE, 77
Usiskin, Zalman, 198

van der Poorten, Alf, 47
Van Slyke, Andy, 10
Vanishing Identity, 161, 163, 177
Vector
    addition, 206
    multidimensional, 3
    space, 151
Vertex (of a graph), 130, 150
Vizing's Theorem, 310
Vizing, V. G., 154
Voles, Roger, 21

Walser, Hans, x, 172, 181, 197, 198, 269
Water, 96
Webb, John, 21
Wedges off the edges, 195
Weight, 172
Well-dressed student, 313
Whitehead, Tamsen, 228, 233
Wild goose chase, 15
Wilde, Oscar, 2
Wiles, Andrew, viii, 23, 33, 40, 42, 44, 45
Willoughby, Stephen, 269
Wilson, P. D. C., 233
Wilson, R. J., 157, 158
Witten, Edward, 235
Witten theory, 235
Wolfskehl, Paul, 39
Wolfskehl Prize, 45
Woman mathematician, 16
Wylie, Shaun, 198

Young student, 127

Zero sextant, 169

Undergraduate Texts in Mathematics

(continued from page ii)

Halmos: Finite-Dimensional Vector Spaces. Second edition.

Halmos: Naive Set Theory.

Hämmerlin/Hoffmann: Numerical Mathematics. *Readings in Mathematics.*

Harris/Hirst/Mossinghoff: Combinatorics and Graph Theory.

Hartshorne: Geometry: Euclid and Beyond.

Hijab: Introduction to Calculus and Classical Analysis.

Hilton/Holton/Pedersen: Mathematical Reflections: In a Room with Many Mirrors.

Hilton/Holton/Pedersen: Mathematical Vistas: From a Room with Many Windows.

Iooss/Joseph: Elementary Stability and Bifurcation Theory. Second edition.

Isaac: The Pleasures of Probability. *Readings in Mathematics.*

James: Topological and Uniform Spaces.

Jänich: Linear Algebra.

Jänich: Topology.

Jänich: Vector Analysis.

Kemeny/Snell: Finite Markov Chains.

Kinsey: Topology of Surfaces.

Klambauer: Aspects of Calculus.

Lang: A First Course in Calculus. Fifth edition.

Lang: Calculus of Several Variables. Third edition.

Lang: Introduction to Linear Algebra. Second edition.

Lang: Linear Algebra. Third edition.

Lang: Short Calculus: The Original Edition of "A First Course in Calculus."

Lang: Undergraduate Algebra. Second edition.

Lang: Undergraduate Analysis.

Lax/Burstein/Lax: Calculus with Applications and Computing. Volume 1.

LeCuyer: College Mathematics with APL.

Lidl/Pilz: Applied Abstract Algebra. Second edition.

Logan: Applied Partial Differential Equations.

Macki-Strauss: Introduction to Optimal Control Theory.

Malitz: Introduction to Mathematical Logic.

Marsden/Weinstein: Calculus I, II, III. Second edition.

Martin: Counting: The Art of Enumerative Combinatorics.

Martin: The Foundations of Geometry and the Non-Euclidean Plane.

Martin: Geometric Constructions.

Martin: Transformation Geometry: An Introduction to Symmetry.

Millman/Parker: Geometry: A Metric Approach with Models. Second edition.

Moschovakis: Notes on Set Theory.

Owen: A First Course in the Mathematical Foundations of Thermodynamics.

Palka: An Introduction to Complex Function Theory.

Pedrick: A First Course in Analysis.

Peressini/Sullivan/Uhl: The Mathematics of Nonlinear Programming.

Prenowitz/Jantosciak: Join Geometries.

Priestley: Calculus: A Liberal Art. Second edition.

Protter/Morrey: A First Course in Real Analysis. Second edition.

Protter/Morrey: Intermediate Calculus. Second edition.

Roman: An Introduction to Coding and Information Theory.

Ross: Elementary Analysis: The Theory of Calculus.

Samuel: Projective Geometry. *Readings in Mathematics.*

Saxe: Beginning Functional Analysis

Undergraduate Texts in Mathematics

Scharlau/Opolka: From Fermat to Minkowski.

Schiff: The Laplace Transform: Theory and Applications.

Sethuraman: Rings, Fields, and Vector Spaces: An Approach to Geometric Constructability.

Sigler: Algebra.

Silverman/Tate: Rational Points on Elliptic Curves.

Simmonds: A Brief on Tensor Analysis. Second edition.

Singer: Geometry: Plane and Fancy.

Singer/Thorpe: Lecture Notes on Elementary Topology and Geometry.

Smith: Linear Algebra. Third edition.

Smith: Primer of Modern Analysis. Second edition.

Stanton/White: Constructive Combinatorics.

Stillwell: Elements of Algebra: Geometry, Numbers, Equations.

Stillwell: Mathematics and Its History. Second edition.

Stillwell: Numbers and Geometry. *Readings in Mathematics.*

Strayer: Linear Programming and Its Applications.

Toth: Glimpses of Algebra and Geometry. Second Edition. *Readings in Mathematics.*

Troutman: Variational Calculus and Optimal Control. Second edition.

Valenza: Linear Algebra: An Introduction to Abstract Mathematics.

Whyburn/Duda: Dynamic Topology.

Wilson: Much Ado About Calculus.